*Cartographic Fictions*

Cartographic Fictions

MAPS, RACE, AND IDENTITY

KAREN PIPER

RUTGERS UNIVERSITY PRESS
New Brunswick, New Jersey, and London

LIBRARY OF CONGRESS CATALOGING-IN-PUBLICATION DATA

Piper, Karen Lynnea, 1965–
    Cartographic fictions : maps, race, and identity / Karen Piper.
    p. cm.
    Includes bibliographical references and index.
    ISBN 0-8135-3072-5 (cloth : alk. paper) — ISBN 0-8135-3073-3
(pbk. : alk. paper)
    1. Cartography—Social aspects.   2. Longitude—Prime meridian.
3. Aerial photography in geography.   4. Geographic information systems.
I. Title.
    GA108.7.P56 2002
    912—dc21   2001048792

British Cataloging-in-Publication information is available
from the British Library.

Contents

ILLUSTRATIONS

PREFACE

As THIS BOOK was going to press, the United States entered what has been called the "first war" of the twenty-first century. President George Bush has said that this war will be different from other wars. It will be a sustained effort against terrorism, until terrorism is eradicated around the world. This war will not honor borders in its quest for terrorists. Now, already, we speak of "cells" rather than nations, and our civil liberties may be sacrificed in order to detect the cells within. George Bush has established a "Home Security Council" and supported an "anti-terrorism" bill that would allow for more electronic surveillance in the United States. Representative Richard Armey commented on this bill: "A lot of this deals with technical questions about your capacity to keep track of 'Joe Bad Guy' without snooping on me, too." Bush promises that the war against "Joe Bad Guy" will last, at least, through his tenure; he has said that he will "hunt down" the terrorists, wherever they are.

The distinction between war and everyday life appears to be collapsing as we enter the twenty-first century. The threat to the United States can no longer be located in any one country but is potentially everywhere. Arundhati Roy, in the 29 September 2001 issue of the *Guardian,* commented: "Terrorism has no country. It's transnational, as global an enterprise as Coke or Pepsi or Nike." Because the terrorists are highly mobile, this war may expand endlessly, as more countries are found to be "harboring" terrorists, including the United States. The war against terrorism may be, in the end, a war against ourselves. Michael Ondaatje, in *The English Patient,* describes a moment when the main character loses track of the enemy during World War II. Ondaatje explains: "He could have been, for all he knew, the enemy he had been fighting from the air." In a similar fashion, we have lost track of the enemy in the United States. Since the collapse of the World Trade Center, the United States has invested all its energy into trying to sort out who the enemy is, in order to push him or her outside our borders. We have been spending billions of dollars in an attempt to reassure ourselves that the enemy is not within. The war against terrorism has been, above all, a reason to increase surveillance, to throw money into defense, to develop new military technologies and mass-produce old ones. There really is no war; there is simply, finally, an acknowledgement that war has become a way of life. This book charts this shift to the notion that war is a sustained way of seeing and mapping the world, which gradually occurred over the last century. George Bush now promises that the United States will "smoke the enemy out of its cave." It will shine its floodlight-satellites into every corner of the world, disassembling nations in search of terrorists.

The United States is now trying to reestablish borders that no longer exist—or to push the enemy *out,* back into the margins of the maps. The history of cartography has similarly been a history of coding the enemy, making a "them" and "us" that can be defended with a clear border. It has been, above all, a history of pushing "them" out of territory that is considered ours—denying their existence, deleting their maps, drawing lines in the sand. But, as I argue in this book, the enemy has always been us, and so the project has been destined to fail, forced to reinvent itself again and again in search of better and more elaborate methods of detection. The United States, after all, was giving arms to Osama Bin Laden long before we decided to attack Afghanistan. This scenario has repeated itself throughout the world, from the Manuel Noriega to Saddam Hussein. The United States, very literally, has been battling itself for a long time. In a less literal sense, the battle of detection has also become a battle against ourselves. "Joe Bad Guy" cannot be located without "snooping on us, too." Joe Bad Guy, after all, may be one of us. Thankfully, satellites with one-meter resolution can see him in his bedroom; they can put terrorists back on the map. Countries who do not want U.S. satellites staring into their houses can now be labeled havens for terrorists, and thus enemies. The borders of the twenty-first century will not be national; they will be closed bedroom doors, they will be caves, they will be "havens."

The maps of the twenty-first century, similarly, will seek to define these ever-shifting borders. Geographic Information Systems (GIS)—a form of computer-based cartography—can create maps that are adaptable, open-ended, and above all smart. GIS can make decisions, predict the future, and effectively persuade an audience. It can store vast amounts of spatial information and be revised constantly to reflect changing conditions. In the Los Angeles Police Department, for instance, a GIS "gang-tracking" program is being used that can record and predict gang movements throughout Los Angeles. GIS can track and map movement—from wildfires to gangs to migrating animals—in a way that paper maps could not. The maps of the new century will be mobile and invasive; their project will be perpetually to redefine the borders of a world that—after September 11—appears to be out of control. The tragedy is that this project is also a war that, it appears, has no end.

Acknowledgments

I WANT TO THANK Derek Martin, especially, for working as my GIS consultant and editor. He read numerous drafts of this book and contributed many of the ideas on GIS. Roland Greene was an inspiring mentor, both professionally and intellectually, through years of work on this project. Deborah Madsen also supported me as a protégé throughout the years, without ever meeting me. I was also supported by the University of Missouri–Columbia Research Council and Research Board with a leave of absence and stipend to travel to London for archival work. In London, the Royal Geographical Society (RGS) let me in the front door, even without the proper papers. They welcomed me as a fellow and kept me well fed during my stay. Huw Thomas, the archivist at the RGS, pulled out the right dusty boxes for me and helped me to sift through them; he also stayed late to let me keep working. The RGS picture librarian, Clive Coward, tracked down the right photos and maps, supplied with only the vaguest of clues from across the Atlantic. At the Royal Greenwich Observatory in London, the archivist, Adam Perkins, patiently helped me to reconstruct the observatory bombing. (The staff of the observatory—from the museum curator to the ticket taker—were extremely friendly and helpful.) Govern Software, Questar, LizardTech, Intergraph, and Orbimage generously donated their advertisements. The Huntington Library provided a beautiful setting for completing the final revision on this book.

Denis Cosgrove introduced me to the field of cultural geography and was as humble and kind as he was brilliant. Winona LaDuke taught me how to write about indigenous cultures and provided, above all, the inspiration for this book. Zsolt Török, thanks to his extensive research and translation work, supplied me with a wealth of information on Lázló Almásy. Jeff Williams, Carsten Strathausen, Noah Heringman, Andy Hoberek, Nancy West, and Elaine Lawless gave me precious editorial advice. (Bobbe Needham caught my mistakes.) My housemate, Robin Albee, endured me through the early stages of this book and proved to be an endless source of geographic information. Finally, I want to thank Mary Piper, who carefully proofed the final draft; though she claimed not to understand it all, she was proud, which was enough.

 *Cartographic Fictions*

Cartographic Cyborgs

IN JUNE OF 1998, I was making my way across London to the Royal Greenwich Observatory, the home of the prime meridian and self-proclaimed "centre of time and space," when I became lost. I was re-reading my map to the London Underground when I noticed an advertisement on the wall that read: "The problem with ordinary maps is, they don't know where you are." This advertisement, produced by Garmin for its hand-held Global Positioning Systems (GPS), seemed to cater to people—like myself—who had gotten off at the wrong subway stop. Instead of an ordinary map, Garmin's GPS could locate my absolute geographical coordinates, via satellite, with the push of a button. It had sentience, even agency, the ad seemed to say. It knew where I was. Once my position had been fixed and my destination entered into this cell-phone-sized computer system, an arrow would come up on the screen that would point me in the direction to walk and even provide me with a map of my route.

The era of the Rand McNally road map, while by no means over, is gradually becoming superseded by an era of cyborg mapping technologies, or technologies that enable the map to think for us. Manfred Clynes, the aerospace scientist who coined the word "cyborg" in 1960, claimed that the purpose of a cyborg is "to provide an organizational system in which ... robot-like problems are taken care of automatically and unconsciously, leaving man free to explore, to create, to think, and to feel."[1] Garmin, in its satellite-based mapping computer, takes care of the problem of being lost by easily locating its own position and consequently that of the holder. "We'll take you there," Garmin's StreetPilot promises. Programs like StreetPilot have been incorporated into the luxury-auto industry in the form of computerized dashboard maps—such as Oldsmobile's GuideStar. The driver merely enters a destination and the car signals when it is time to exit the freeway or turn. 911 uses a Geographic Information System (GIS) to decide the fastest route for

ambulance drivers to travel, eliminating the possibility of human navigational error and allowing the driver to concentrate on driving. Tractors
can now drive themselves using GIS/GPS technology. MapQuest, a
web-based GIS, gives directions, turn by turn, for getting anywhere in
Canada or the U.S. Maps are being designed to think for us, so that we
have more time "to create, to think, to feel" without worrying about
being lost.

Once, while hiking in the Rocky Mountains, I asked someone coming the other direction how far he thought it was to the lake. He
answered, without hesitation, "My GPS says it's 5.8 miles." He was carrying a hand-held Garmin system, and all he had to do was glance down
at his satellite-generated map to know exactly where he was. This kind
of symbiotic relationship between satellites and humans is actively being
promoted in advertisements for GIS/GPS. For instance, in an advertisement that reads "We're Putting on a New Face," the face is the map. The
caption reads: "Technology has gone beyond AM/FM/GIS [the old
mapping system] and so have you," making an equation not only
between the customer and the map but also between two maturing sys-

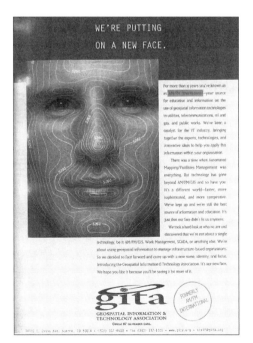

1. The subjects in GIS ads are
predominantly white men.
"We're putting on a new face,"
1998, courtesy of Geospatial
Information & Technology
Association (GITA).

2. This ad demonstrates the cyborg status of GIS. It is unusual in that the map-man is multicolored and almost monstrous looking; a certain slave status seems to be implied by equating "your GIS" with "your man." "Run your GIS to win!" 1998, courtesy of Intergraph Corporation.

tems (fig. 1). New GIS/GPS technologies are constantly superceding each other, leading to better resolution, speed, and user-friendliness. In another advertisement, "Run Your GIS to Win," the caption reads, "In the race for productivity, it takes a fast, open, complete GIS to win. Run with the clear winner." The men who are racing in the illustration have been taken over by their GIS programs. Both ads suggest that getting the newest GIS program will improve one's own performance, making you better and faster (fig. 2).

Not only is GIS being linked to improving human performance, but also mapping programs are being sold for their ability to process vast amounts of global information (or data), making it useful to the individual. Advertisements, therefore, commonly depict mapping data as literally being ingested into the body; satellite photos of the globe are often being carried, thrown, or even eaten. In *Earth Observation Magazine,* a recent advertisement read: "The Larger Your Appetite for GIS Data, The More you Need SQS [a GIS program]." Depicting the globe on a platter, this advertisement suggests that the planet can literally be eaten and in this way controlled. The hand holding the globe on an ornate silver platter is wearing a white glove, invoking a certain nostalgia for the days of the wealthy aristocracy. This ad, which ran in a British magazine, did not appear in the United States, where one is more likely to encounter images like "Mr. SID: The Most Powerful Image Compressor on the Planet" (fig. 3). Rather than a wealthy patrician, a stylish businessman controls the planet in this ad.[2] Another U.S. advertisement depicts the globe as a pizza, reading: "Have Your Slice Delivered" (fig. 4). But what all these corporations are selling (both in the United States and Britain) is a body wed to the map, improved and nourished by the consumption of data.

The question, then, is whose body is being linked to the map and who is given the power to consume and process data. Production of GIS technologies occurs predominantly in first-world countries, just as satellites are generally owned by the most industrialized nations. Joseba Gabilondo wrote, "In the economically privileged First World the production of 'Man' has given way to the reproduction and simulation of

3. Mr. SID promises that "massive images of unlimited size" (including satellite imagery) can be instantaneously made available to zoom into, navigate, and manipulate. "Introducing Mr. SID," 1998, courtesy of LizardTech.

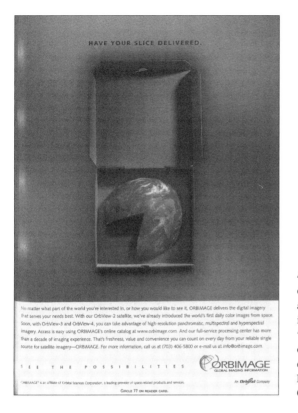

4. Consumption of global data is a common theme in GIS ads. "Have your slice delivered," 1998, courtesy of OrbView Imagery, provided by ORBIMAGE.

'cyborgs.'"[3] But if the first world now produces cyborgs, or people wed to high-tech production and consumption, its demand for data-food has become focused largely on the third world, which is seen as lacking in geographic information. Maps in third-world countries are often out of date, resources are uncharted, and census data are generally weak or unavailable. "The first law of geographical information," one GIS critic suggested, is "the poorer the country, the less and worse the data."[4] Those nations with "less and worse" geographic information, then, automatically become subjects for satellite data acquisition from first-world countries. S.F.H. Borley explains: "Developing countries may lack the data they require, or this data may be out of date and inaccurate. . . . GIS designers have turned to satellite remote sensing and aerial photography to provide a new data source."[5] The collection of data by an extra-national agency, however, sets up a relation that is potentially oppressive, as one GIS critic explained: "It leads experts to see those people to whom their data refer as 'other'. . . . Because the availability of information is seen as being of fundamental importance to the making of decisions, those who have the information see themselves as empirically better able to make decisions than those who are merely 'other.'"[6] The hand that wears the glove in the British GIS advertisement, then, could be read as the third world offering up its data for first-world consumption. The third world feeds the data that make the map man whole, represented, in these ads, with the satellite image of the globe swallowed by the first world.

The contemporary lack of third-world data, in many ways, simply represents a continuation of the constant colonial struggle to fill voids in maps; the failure of maps to adequately cover the entire territory has always been a driving force of cartographic pursuits. As in Jorge Luis Borges's famous parable, the goal in cartography appears to be that the "imperial map . . . ends up exactly covering the territory."[7] Historically, those areas that were outside of geographic knowledge—or off the map—were seen as the abodes of monsters and brutes. In sixteenth- and seventeenth-century colonial maps, monsters regularly inhabited the margins, showing the frightening limits of European knowledge. Geographers in the eighteenth century commonly traced the word "cannibals" over blank spaces on maps. In 1733, Jonathan Swift observed, "So geographers in Afric maps, with savage pictures fill their gaps."[8] The Arctic, also, was depicted as full of "brutes with neither language nor reason [who] hiss like geese."[9] Anne McClintock commented on this practice: "With the word cannibal, cartographers attempted to ward off the threat of the unknown by naming it, while at the same time confessing a dread that the unknown might literally rise up and devour the whole." She continued, "The failure of European knowledge appears in the margins and gaps of these maps in the forms of cannibals, mermaids and monsters, threshold figures eloquent of the resurgent relations between gender, race and imperialism."[10] By moving dangerous or potentially resurgent elements to the marginal spaces on the map, fear of threats to the imperial map were circumscribed. Michel Foucault, in *The Order of Things* and *The Archaeology of Knowledge,* discusses the fear of these kinds of marginal or "threshold" spaces. Because of this fear, he proposes, Europeans constantly attempt "to master and control the great proliferation of discourse, in such a way as to relieve its richness of its most dangerous elements; to organise its disorder so as to skate round its most uncontrollable aspects."[11] Maps, it seems, have been so organized, to skate around danger and delimit the boundaries of knowledge; dangerous elements, in turn, are forced into the blank spaces, oceans, or margins of the maps.

If, territorially, there has been a historical tension between areas that are mapped and unmapped, so, psychologically, it could be said that there is a parallel tension between what is considered human and what monstrous. Jean-Paul Sartre, along with others, has suggested that construction of European "Man," with his notions of "freedom" and agency, is based upon the constitution of "Others." In the preface to *The Wretched of the Earth,* Sartre sums this theory up clearly: "There is nothing more consistent than a racist humanism since the European has only been able to become a man through creating slaves and monsters."[12] In the West, identity is based in the notion of the free and independent individual who is sovereign or self-governing; yet, as many have argued, this concept of the individual emerged concurrently with the growth of the slave trade. Literally, "free" humans could be seen as the inverse correlative to the slave or the "uncivilized" human being. Sovereignty, in European discourse, is linked to establishing control over that which is considered disordered or proliferating, whether it is indigenous peoples or the natural environment.

Mary Louise Pratt has suggested that sovereignty, in colonial discourse, is constructed through sight; this is what she calls the "monarch of all I survey" trope.[13] What is seen is claimed, or thought to be owned; thus, one is sovereign over that which one sees. This was commonly the attitude of explorers, in the race to map the colonies. For instance, when Henry Stanley entered the Lake Tanganyika district of Africa in 1871, he wrote in his journal: "Think how well a score of pretty cottages would look instead of those thorn clumps and gum trees! Fancy this lovely village teeming with herds of cattle and fields of corn, spreading to the left and right of this stream! How much better would such a state become this valley."[14] Today, through "visualization" programs, this same process can be enacted by computers, which redesign the landscape according to the designer's plans. *GIS World* recently devoted an issue to visualization programs, the cover of which read: "Visualize Realistic Landscapes: 3-D Modeling Helps GIS Users Envision Natural Resources." Visualization technologies are combining with GIS programs to help users "see" their product, which in this case is "natural resources." The theory behind visualization technologies is that, for instance, a "virtual" Bermuda could actually be sold to Bermuda as a model for planning. In an interview for *GIS World,* the designer of the Virtual Bermuda project commented that a 3-D representation of Bermuda would help to establish and control definitions of "the Bermuda image."[15] In this example, Bermuda was completely relandscaped, and the poor and homeless were eliminated from the "map." So while Stanley could enter Africa and see England, now computer-mapping technologies can visualize virtual Bermudas. Just as Stanley could see pretty cottages rather than thorn clumps, it is as if resources may be finally envisioned—or brought into being—by the power of the gaze alone. Thorn clumps are seen, but pretty cottages establish sovereignty.

Sovereignty, since the eighteenth century, has been defined in international law as "the control of a well-defined territory"; and "territory" tautologically meant the land "under the jurisdiction" of a sovereign.[16] Legally, a sovereign state could acquire territory through "an act of effective apprehension, such as occupation or conquest."[17] The way to establish sovereignty was to mark a boundary or make a map, a method accepted in international law. The law stated that "states may by convention fix limits to their own sovereignty, even in regions such as the interior of scarcely explored continents where such sovereignty is scarcely manifested, and in this way each may prevent the other from any penetration of its territory."[18] Cartography, then, became a race to imprint the "scarcely manifested" record of sovereignty upon a territory.

Preconquest territories, according to this definition, belonged to no one. In Australia, the doctrine of *Terra nullius,* held until 1992, defined preconquest Australia as a "territory belonging to no state, that is, territory not inhabited by a community with a social and political organisation."[19] To be sovereign, then, involved taking land from those who were considered less "organized."[20] It was based in the idea of invading a void, or an unoccupied space, which—of course—existed nowhere but in the colonial imagination. Sovereignty became a way to rhetorically clear space for invasion, and in this clearance, the concept of whiteness—as transparency—could emerge. Geographer Robert Sack explained this phenomenon: "Territoriality in fact creates the idea of a socially *empty* space."[21] The modern conception of space involves a perpetual *separation* of places and things followed by their recombination "as an assignment of things to places."[22] Thus, we have the notion of "virgin" or "empty" land that is waiting to be filled. Sovereignty, in this sense, became linked to erasure, based in the notion of creating a territorial blank slate on which one could construct colonial rule and authority. Clearing space, in effect, became a way to establish whiteness—or to differentiate oneself from indigenous peoples.

If the sovereign state could acquire territory through occupation, so could the sovereign individual. Thus the notion of private property emerged in conjunction with the idea of individual sovereignty. In *Robinson Crusoe,* the character of Crusoe exemplifies the ideal of gaining individual sovereignty through acquiring territory. Crusoe, after years of cultivating the island that he was stranded upon, declared with satisfaction: "I was the lord of the whole manor; or if I pleased, I might call myself king, or emperor over the whole country which I had possession of. . . . I had no competitor, none to dispute sovereignty or command with me."[23] As a kind of mini-kingdom, Crusoe's island is seen as his even after he leaves it, simply because he had claimed it and cultivated it before any other European. Jean Jacques Rousseau, in *The Social Contract,* also spoke of the rights of "first occupancy" over a territory: "In order that the right of 'first occupancy' may be legalized, the following conditions must be present. (1) There must be no one already living on the land in question. (2) A man must occupy only so much of it as is necessary for his subsistence. (3) He must take possession of it, not by empty ceremony, but by virtue of his intention to work and to cultivate it."[24] This philosophy of private property could be read as an extension of the colonial methods of acquiring territory, in which the individual establishes his or her dominance over nature/natives. In *Robinson Crusoe,* Friday cannot claim possession of the island, even though he works on it; the "cannibals" who regularly use the island for their ceremonies also have no rights to it.

Private property, it was believed, extended from one's inalienable right over his or her body. Then, what one did with the body (i.e., labor) also became private property. John Locke, in his *Second Treatise of Government,* made this progression explicit, stating: "Every man has a property in his own person; this nobody has any right to but himself. The labor of his body and the work of his hands, we may say, are properly his. Whatsoever, then, he removes out of the state that nature has provided and left it in, he has mixed his labor with it, and joined to it something that is his own, and thereby makes it his property."[25] Removing something from the state of nature meant establishing sovereignty; and so "nature" itself, as well as the indigenous peoples that resided within it, was seen as an obstacle that must be overcome. The land became "civilized," literally, when it became an extension of the European's body. The fear was that those who had been cast out of the body/land, like some demonic ghosts, would return.

Because there was always an awareness of those who were pushed aside in order to construct whiteness and clear territory, a fear of the "primitive" concurrently would emerge. The term "primitive," not surprisingly, became popularized during the eighteenth century; previously, the word "savages" had been commonly used.[26] The main difference between these two terms was that "primitive" linked aboriginal peoples to the idea of an originary moment; "savages," in contrast, did not have temporal associations. Hadyen White explains that, by the nineteenth century, "primitive man" came to be regarded "as an example of arrested humanity, as that part of the species which had failed to raise itself above dependency upon nature."[27] "Primitive" signified the inverse of "progress" and "development," and the term would gradually begin to appear everywhere, from Marx's primitive food gatherers to Freud's primal horde and Nietzsche's barbarians.[28]

Because of their presumed disorganization and arrested development, "primitive" cultures were viewed by the colonizers as infected with fear, superstition, enchantment, or fancy. "Enlightenment," according to Max Horkheimer and Theodor Adorno, "has always aimed at liberating men from fear and establishing their sovereignty." In *Dialectic of Enlightenment,* they describe this phenomenon: "The program of the Enlightenment was the disenchantment of the world; the dissolution of myths and the substitution of knowledge for fancy."[29] The emergence of the concept of sovereignty became invested in pushing out fear from the territory, thus overcoming its primitive disorganization. The etymology of the term "territory" also suggests this relation between fear and sovereignty. According to the *Oxford English Dictionary,* the term "territory" is derived—in the fifteenth century—from the French *terrere,* or to frighten (terrorize), and *territor,* or frightener (terrorist). The etymologi-

cal origins of "territory" were later displaced—in the eighteenth century—by the Latin *territorium*, which combines *terra* ("land") and *torium*
("belonging to" or "surrounding"). "Territory," it seemed, was something
haunted from within by the "primitive," which had to be perpetually
overcome by the sovereign subject.[30] The "primitive" was what the sovereign subject hoped to displace, in his or her role as controller or organ-
izer of space. Horkheimer and Adorno explained that any reversion in
this progressive dialectic was thought to involve "a reversion of the self
to that mere state of nature from which it had estranged itself with so
huge an effort, and which therefore struck terror into itself."[31] The terror associated with primitive life and organization became associated
with a fear of reversion, as well as invasion.

Sigmund Freud later capitalized upon this idea of reversion—or the
primitive within—in the field of psychology. Freud claimed that individual development paralleled anthropological stages of development,
suggesting that "primitive beliefs are most intimately connected with
infantile complexes, and are, in fact, based upon them."[32] The "civilized"
individual, he suggested, had the ability to overcome his or her primitive
complexes in ways that aboriginal cultures had never succeeded in doing.
Freud's "uncanny" was that forgotten history that threatened to pull the
civilized back into a primitive state: "The 'uncanny' is that class of the
terrifying which leads back to something long known to us, once very
familiar."[33] The uncanny was described as that which was once familiar
but had been forgotten; it was strangely "home-like" but nonetheless
frightening. The problematic pull between the familiar (home) and the
primitive created the feeling of the uncanny, according to Freud, which
led back to "the old, animistic conception of the universe" in which "the
world was peopled with spirits of human beings."[34] Freud continued,
"Nowadays . . . we have *surmounted* such ways of thought; but we do not
feel quite sure of our new set of beliefs, and the old ones still exist within
us ready to seize upon any confirmation. As soon as something actually
happens in our lives which seems to support the old, discarded beliefs we
get a feeling of the uncanny."[35] The uncanny captures that fear of the
"primitive" as an object that cannot be superseded. It is frightening precisely because the "primitive" is *us*. It is that part of ourselves that we
thought had been surmounted, but that was actually contained within
our origins, as *home,* waiting to be remembered.

The idea of the map was invested in overcoming the darkness of primitive territorial organization and establishing sovereignty, as whiteness, as home; but it left residues of the uncanny within its borders. The colonial map, in this sense, could be said to be as much about the boundaries of modern human identity as it is about territorial designations. It is about overcoming the possibility of primitive knowledge, both about the land and within the self. Ironically, of course, explorers could only create maps by relying upon indigenous geographic information and native guides to the territory. In 1874, the president of the Royal Geographical Society described the mapping of northern India as a threefold process: "First . . . were the reports of native travellers which shed a wide but uncertain light on the vast unknown. Behind them, piercing this gloom, came narrow shoots of clear light representing the travels of individual European explorers. Finally, and well behind, came the zone of harsh white reality shed by the surveyors and map-makers."[36] In colonial discourse, the natives are often described as having insufficient or "uncertain" knowledge, thus justifying the entrance of the colonizer who, by contrast, appears to have certain, unambiguous, and scientific knowledge. In 1818, one explorer commented on the lack of indigenous knowledge about their own territories: "Do the savages of New Holland, we would ask,—do the Hottentots of the Cape—do the more civilized tribes of African negroes, or of the Eskimaux of Greenland—do any one of these know the extent of their respective countries?"[37] The answer to this rhetorical question would provide the justification for further cartographic pursuits. The insufficiency of native knowledge—its very primitivism—became the defense for invasive mapping projects.

Even as the cartographer attempted to overcode native knowledge, he was generally reliant upon indigenous peoples for their knowledge of the territory. Therefore, the cartographer's job was always an ambivalent one. The cartographer had to rely upon the so-called "native informant" who generally provided the information for the map, but who could also utilize any number of strategies to confuse or resist the cartographic projects. For instance, in nineteenth-century India, locals were known to pull up the stakes that marked the surveyor's base line for triangulation surveys. In one explorer's report of this situation, he explained: "In India such marks are viewed with cupidity not unmixed with fear. The natives have an idea that money is buried under these mysterious monuments erected by the western strangers, while they feel a dread that they may cast a spell over the district."[38] By turning the native's understandable fear of territorial encroachment into a superstitious dread, the colonizer thus attempted to belittle or stereotype this threat, thus containing it. But, overall, the cartographer's distrustful relationship with the native informant would lead to the desire to eliminate this source of knowledge.

Cartography, then, began moving away from the "threat" of the native, often literally by leaving the ground. The development of the aerial camera in 1915 signaled this possibility, which cartographers openly celebrated: "Mappers were no longer required to 'slog' into the messy reality of the field in order to produce the 'map.' . . . The need for field survey and the actual contact of the cartographer with the object of his or her work was, as a consequence, greatly reduced."[39] Avoiding the "messiness" of reality, the cartographer could also avoid the unreliability of the native, who was seen as a more literal threat to mapping projects. Similarly, the camera was seen as aiding in evading the "spell" of the native—or his possible ambush. One advocate of aerial photography explained: "The surveyor is no longer travelling blindly, exploring as he goes, nor is he dependent on Indian guides to lead him on his route."[40] The ambivalent relation with the "Indian guide" can be seen in this quote; traveling "blindly" is associated with dependence upon the Indian. By contrast, "seeing" (or regaining sight) can be achieved with the aid of a plane, which eliminates the Indian. In this sense, the elimination of the native is dependent upon the production of the pilot who "sees" through the camera. Similarly, the "less and worse" data of third-world countries may today be overcome through the first-world satellite.

In this book, I read particular shifts in mapping technology—the establishment of the prime meridian, the development of aerial photography, and the emergence of satellite/computer mapping—as representative of cartography's impulse to leave the ground in order to escape the dangerously racialized or gendered subject. This view from above has been the false trajectory of cartography, which seeks to move into space in order to overcome race. Maps, of course, are generally not thought of in terms of race. Maps are still largely read in a utilitarian fashion—to get around. In this sense, they are understood as asexual and deracialized objects of territorial information—which further promotes their claims to objectivity. In order to fulfill its fantasies of objectivity, colonial discourse eliminates the very indigenous knowledge upon which it relied to produce the map.

My reading of cartography is in direct contrast to those in traditional histories of geography and cartography, which generally proceed as narratives of accretion in which "man" is the subject and the knowledge of the earth is the object. Generally, cartographic knowledge is presumed to gradually grow as frontiers are pushed forward until the map occupies all. Instead, I argue that cartography is equally invested in constructing "man." Cartography, in my reading, is part of a colonial discourse invested in establishing "whiteness," or transparency, as a kind of identity formation. This is not to say, however, that cartography is simply the history of domination over indigenous peoples, in an effort to create colonial authority. The history of colonial cartography certainly contains this violence (including direct military action), but the dominating discourse never quite succeeds. It never quite suppresses alternative forms of territoriality, which continue to haunt the map. Similarly, the progress of the map, itself, could be read as a kind of cognitive failure—or a form of mistaken identity.

The contest over territorial definitions still occurs, though the very struggle may be misunderstood. Like the struggle over the "official" or state-sponsored language, cartography is based in the contest over whose map wins official status. To illustrate this point, a student from Zambia wrote of her admiration for non-English-speaking women in rural areas who were "quick-witted, intelligent, very skilled socially, and . . . very nimble and agile conversationalists." This student saw these women treated "by the official or bureaucratic world" as "illiterate peasant women."[41] This transformation from "agile conversationalist" to "illiterate peasant woman" demonstrates the struggle over representation that also occurs in territorial designations. Some maps are seen as credible, or more sophisticated, and so are given state sponsorship and become the official territorial language; others are misunderstood, misrepresented, and forgotten. But they still continue. It is this gap in cartographic knowledge—as well as the complex ways in which this very gap is avoided, overcoded, or suppressed—that is the subject of this book.

I begin this book in Greenwich, England, because this is where the concept of global space and time came into being. In 1884, the establishment of an internationally accepted prime meridian (or 0° longitude) at Greenwich marked the transformation of cartography into an international discourse in which a "universal day" was accepted as well as a basis for standardized map-making around the world. Patrick McHaffie explained that the establishment of this global grid was essentially an attempt to delimit territorial meaning: "By basing the subdivision of space on a worldwide grid such as latitude and longitude, these mapping systems tear local meaning from areas."[42] Greenwich represented an attempt to eliminate competing meridians in Lisbon, Rio de Janeiro, Paris, and elsewhere. Chapter 1 demonstrates how anxieties about multiplying meridians—a form of proliferative discourse—led to the establishment of one prime meridian in Greenwich. This elimination of meridians, interestingly, paralleled a fear of "polyglot discourses" invading England through waves of anarchist immigration from Russia and France. Joseph Conrad's *The Secret Agent* documents this fear of anarchism through the retelling of an actual anarchist attempt to blow up the Greenwich Observatory. Newspaper accounts and police reports from this time period also reveal that the concept of "Greenwich time" was thought to be under attack by the suffragettes, who aligned themselves with anti-imperialist movements. It is in this in-between space—between the suffragettes and Greenwich—that my reading takes place.

Beginning in Greenwich, I go on to examine the triangulation surveys—based on latitude and longitude—that occurred in the Himalayas, looking specifically at the dialogue that occurred between the "pundits" and explorers in mapping Tibet. In the Survey of India maps of the Himalayas, the corners of the maps contain a line that reads: "From the Routes of Pundit A_K_." Behind this simple line is a hybrid history of colonial exploration, in which Indians—in disguise—traveled into Tibet to map territory for the British. The ambivalent relationship between the European officers and these "Pundits" was recorded in explorer journals and popularized in Rudyard Kipling's *Kim*. Chapter 2 will examine the history surrounding this marginal note, revealing the ambivalent statements about racial identity that it represents. Chapter 3 investigates the rise of aerial photography in the 1920s and women's role in its development. Chapter 4 examines the postcolonial reformulation of the Nazi mapping of Africa in Michael Ondaatje's *The English Patient,* looking at journals of the actual explorers and bringing back narratives of resistance by the Libyan Senussi. Chapter 5 compares the Canadian development of GIS to Margaret Atwood's feminist "explorer" narrative, *Surfacing*, as well as the local Cree and Innu resistance to mapping projects in Quebec. Chapter 6 discusses the GIS program used by the International Water Council in Amitav Ghosh's *The Calcutta Chromosome*, examining postcolonial resistance to mapping technology in this text and comparing it to the anti-dam movement in India, specifically looking at the activist writings of Booker Prize–winning novelist Arundhati Roy.

In each of these chapters, I explore the tension between those who map and those who resist or redefine mapping projects. I also look at the psychic conflict within the cartographers themselves, as they struggle to push toward those "blank spaces" that must be mapped. It is in the margins of the map, I will demonstrate, that cartographers and explorers repeatedly describe an uncanny fascination with the primitive. In Joseph Conrad's *Heart of Darkness*, this fascination with the primitive (and its dangers) is popularized in the character of Kurtz, who travels into the "darkness" of Africa and is transformed by it. It is this transformation, on the margins of the map, that is repeatedly described by explorers as both seductive and threatening. To evade the boundaries of official colonialism and *go further* is precisely the fascination of cartographic pursuits.

These margins, or unexplored territories, are described in explorer narratives as both objects of desire and areas that are being lost, or swallowed up, by the map itself. As the boundaries of the map are pushed forward, a certain nostalgia for unmapped spaces begins to emerges. There is also, however, a fear that what is discovered *out there* may escape its margins and be brought back home or somehow infect the civilized world. So we see a predominance of European literary figures like Frankenstein's monster, who shows up in England, or Conrad's Kurtz, who becomes infected—at his moral core—by the "primitive" cultures of Africa. Timothy Findley's *Headhunter*, in its memorable opening lines, describes a fear of Kurtz coming back from Africa: "On a winter's day, while a blizzard raged through the streets of Toronto, Lilah Kemp inadvertently set Kurtz free from page 92 of *Heart of Darkness*. Horror-stricken, she tried to force him back between the covers."[43] The horror of Kurtz potentially coming home is the horror (and desire for) the primitive, who—like Frankenstein's monster—will run amuck, lost, destroying the world. It is the fear of what has been left off the map that makes Kurtz show up in Toronto or monsters appear in the margins.

Western identity is formulated by pushing something off the map, then safely embracing the map as the self; but knowledge of the margins is always waiting to return, as the uncanny.

To be "lost," then, or to be unable to find yourself on the map, is to become caught in a problematic fantasy of identification with that which has been pushed off the map. Michael Ondaatje, in *The English Patient,* describes a moment of being "lost" in London. A group of geographers who just returned from Africa are looking for the Royal Geographical Society. Ondaatje explains, "When they travel by train from the suburbs towards Knightsbridge on their way to Society meetings, they are often lost, tickets misplaced, clinging only to their old maps." They are, he writes, "like Conrad's sailors . . . not too comfortable with the etiquette of taxis, the quick, flat wit of bus conductors."[44] This sentimentalizing of being lost signals a desire to evade the effects of "over-civilization" and so to jump off the official map and into the margins or blank spaces. In this sense, I was also "lost" in the London Underground. I was outside of Garmin, the Tube maps, the Greenwich Meridian. I was also displaced, a foreigner in London, dreaming of home and thinking that I did not belong in the world of taxis and trains. Thankfully, this problematic fantasy of escaping the map—and thus my own cyborg status—did not last long. I soon found myself on the map. It is, after all, nearly impossible to get lost in the London Underground, a train system now run by GIS.

PART ONE

The Prime Meridian

Chapter 1

The Suffragette Disturbances and the Bombing of Greenwich

On October 20, 1905, the militant campaign for women's suffrage began in England when Christabel Pankhurst attempted to "spit" on a police officer. She explained, "It was not a real spit but only, shall we call it, a 'pout,' a dry purse of the mouth."[1] Though Pankhurst's notorious "spit" became only a pout, this pout would be enough to have her thrown in jail, becoming the cause célèbre for militant feminism. Millicent Fawcett claimed that after this incident "the whole country, indeed we might almost say the whole world, rang with the doings of the Suffragettes"—however, Fawcett also noted that "while they suffered extraordinary acts of physical violence, they used none, and all through, from beginning to end of their campaign, they took no life, and shed no blood, either of man or beast."[2] The suffragettes in Britain were known for throwing stones at police cars, breaking windows, planting bombs, vandalizing works of art, pouring acid on golf courses, and cutting telephone wires.

The reason for this violence, some say, was the betrayal of suffragists in 1884. The Franchise Bill that passed in 1884, making suffrage universal for men, deliberately evaded the question of women's suffrage. In fact, the number of Liberals who supported women's suffrage made an astonishing drop from 75 percent in 1883 to 26 percent in June of 1884. Interestingly, by the end of 1884, the numbers had returned to 84 percent in favor of women's suffrage.[3] In 1884, the largest majority of Liberals voted against women's suffrage since the beginning of the movement to enfranchise women.

Brian Harrison explained this sudden change in mind among Liberals as symptomatic of "the disillusionment felt by revisionist utilitarians during an age when (particularly after the Home Rule crisis of 1886) democracy no longer seemed fully compatible with strong and efficient government."[4] This link between the Irish demand for home rule and the women's suffrage movement was not a new one. Though the comparison was initially made by antisuffrage leaders, the suffragists themselves would begin to model their methods on those of the Irish and other colonized nations. In 1911, Pethick Lawrence, in the trial of Christabel Pankhurst, made this analogy explicit by claiming, "If a referendum is to be adopted on Women's Suffrage, it ought to be adopted on Home Rule and Welsh Disestablishment." The press picked up on this analogy, comparing women to the Irish: "What would be the temper of the Nationalist party if the Government had proposed . . . to introduce a Home Rule Bill for England, another for Scotland, another for Wales, and to leave Ireland under the Imperial Government? That was what the Government proposed to do for the women."[5]

This link between nationality or ethnicity and gender may appear strange to a contemporary reader. But, as Anne McClintock argues in *Imperial Leather,* "domestic space became racialized" in the Victorian era. "Imperialism suffused the Victorian cult of domesticity and the historic separation of the public and the private, which took shape around colonialism and the idea of race . . . colonialism took shape around the Victorian invention of domesticity and the idea of the home."[6] The home, in essence, became coded white, representing obsessions with tidiness, chastity, and leisure. In contrast, those who were outside the home, including agitators for suffrage, were increasingly described as outside of British national identity, in comparisons either to the nonwhite colonized or to other national groups perceived as revolutionary. In 1893, George Egerton wrote that there is an "untamed primitive savage temperament that lurks in the mildest, best women."[7] Besides the Irish and the nonwhite colonized, the suffragists were most often compared to the Russians. In the French magazine *Le Rire,* a "feminist" is characterized as a Russian terrorist (fig. 5). Pankhurst claimed: "If you read the leading articles in some of our newspapers, you would think our methods were Russian methods, or even worse. You would really suppose that we were the most dangerous set of people and the most violent set of people, that have ever been seen. The fact is, however, that we are singularly mild." Given Egerton's appraisal of "mild" women, however, this protestation to mildness may not have been enough to alleviate the public's fear, which

5. Suffragettes were often
described as anarchists, an
image depicted here in the
satirical journal *Le Rire* at the
turn of the twentieth century.

Pankhurst aptly described: "I know this—they are more afraid of one
Suffragette than they are of 5,000 men!"[8]

The link between suffragists and non-natives was often embraced by
the suffragettes themselves, who actively adopted the role of extranationals to promote their revolutionary designs. Pankhurst declared in
1913: "Negotiations are over. War is declared."[9] If the cult of domesticity
was intended to keep women in their place as subservient citizens of
male-defined nationality, women who stepped out of that space posed a
threat to notions of national identity and civilization. Anne McClintock
points out that the verb "domesticate" carried as one of its meanings "to
civilize." She concludes: "Through the rituals of domesticity . . . animals,
women and colonized peoples were wrested from their putatively 'natural' yet, ironically, 'unreasonable' state of 'savagery' and inducted through

the domestic progress narrative into a hierarchical relation to white men."[10] The contained woman, it seemed, was essential to the notion of civility upon which the empire was built.

If the suffragettes did not eschew comparisons to the colonized, neither did the resistance movements in the colonies distance themselves from associations with the suffragettes. Mahatma Gandhi's civil disobedience campaign, it has been claimed, was inspired by the suffragette disturbances in the early 1900s.[11] Millicent Fawcett writes of receiving money from an Irish nationalist for the suffragette cause, with a note attached: "Give the money to those good women who are persecuting the Government."[12] Suffragettes were viewed as stepping outside the carefully tended boundaries of home, nation, and even gender. Like Homi Bhabha's "mimic men," the suffragettes, it was believed, wanted to be men; but, as one critic pointed out, "imitations are generally bad and clumsy."[13] Bhabha claimed that a certain horror was evoked by the native who tried to be "white" but was "not quite/not right" in his or her mimicry of the customs of Europeans. Women, in crossing the boundaries of what coded them as women, similarly evoked an uncanny fear in which one suffragette could be deemed more terrifying than five thousand men.

The multiplication of these "women as men" (who may appear mild or normal but are not!) paralleled fears of colonial space being out of control. The defeat of William Woodall's amendment for women's suffrage in June of 1884, strikingly, was followed by the International Meridian Conference in October of 1884, which chose Greenwich, England, as the "centre of time and space" for the world. The conference, held in Washington, D.C., declared that "it is the opinion of this Congress that it is desirable to adopt a single prime meridian for all nations, in place of the multiplicity of initial meridians which now exist."[14] Greenwich was chosen as the "prime meridian of the world," serving to standardize cartographic knowledge and eliminate the perception that cartographic laws were random, disputable, or even political. Fredric Jameson has written of the effects: "We no longer are encumbered with embarrassment of non-simultaneities and non-synchronicities. Everything has reached the same hour on the great clock of development or rationalization."[15] Greenwich had already defined 0° longitude for 72 percent of the world's trading ships, which depended on sea charts using

Greenwich as the prime meridian. The establishment of the prime meridian in Greenwich, however, would permanently sanction Britain's empire as the imperial center of the world.

Before 1884, there were also prime meridians in cities such as Paris, Lisbon, Rio de Janeiro, and Washington. Each city in the world kept its own time, based upon the position of the sun at noon; when the sun was directly overhead, clocks would be set at noon. The advent of the railroad, however, meant that trains could pass through more than twenty times zones in one day. Because of the problem of maintaining accurate time schedules, railroads developed their own time systems, which meant that a passenger would not only have to keep local time but also railroad time. The general confusion and inconvenience of coordinating time between different cities around the world was a major reason for adopting a single prime meridian as a standard for timekeeping. The concept of "Greenwich time," however, was not without its opponents. A letter to the editor of *Blackwood's Magazine* demanded, "What in the name of whitebait have we to do with Greenwich more than Timbuctoo, or Moscow, or Boston, or Astracan, or the capital of the Cannibal Islands?" Even the astronomer royal himself, many believe, was not convinced by the concept of universal time. An anonymous letter to the editor of the *Illustrated London News*—thought to be by the astronomer royal—complained of the observatory's "desperate attempts to 'annihilate both Space and Time'" through the introduction of Greenwich time. He pleaded with the public to allow "the old legitimate King Time to resume his place in our clocks and bosoms." Claiming that the "uniformity" of time was a conspiracy of the "clerks who settle the railway time-tables," this anonymous contributor pled for a return to the laws of nature and the timetable set by the sun.[16] Britain's predominance as a colonial power, however, combined with its massive sea power and development of navigational technology, ultimately led to Greenwich being designated the "prime meridian of the world."[17]

It may prove impossible to locate any direct link between the conference in June of 1884, which disavowed women's suffrage, and the conference in October, which naturalized Greenwich as the center of time and space for the world. However, the drop in Liberal support for suffrage just prior to the Meridian Conference is significant enough to warrant an investigation into the relation between the globalization of space/time and the role of gender. Mary Louise Pratt has suggested that what she calls "planetary consciousness" emerged during the eighteenth century: "For what were the slave trade and the plantation system if not massive experiments in social engineering and discipline, serial production, the systematization of human life, the standardizing of persons?"[18] Though the eighteenth century provided the ideological mold for this type of project, it would be the nineteenth century that provided the technology for globalization. It was at this time, according to McClintock, that "panoptical time came into its own," by which she means "the image of global history consumed—at a glance—in a single spectacle from a point of privileged invisibility." The instruments for creating panoptical time were, according to McClintock, "surveying, map-making, measurement and quantification." The domestic realm, she claimed, "far from being abstracted from the rational market, became an indispensable arena for the creation, nurturance, and embodiment of these values." In this sense, "the cult of industrial rationality and the cult of domesticity formed a crucial but concealed alliance."[19]

In an advertisement for Globe silver polish, this analogy is clearly represented (fig. 6). Dressed in a helmet and British flag, with the imperial image of a rising sun behind her, the well-known figure of Britannia is polishing the globe. In this ad, home and empire come together in an evident union—the woman's role is to clean up the world, suggesting empire's civilizing mission and its relation to the domestic realm as an emblem of civility. The ad also, however, clearly shows the grid of latitude and longitude imposed upon the globe. The visual correlation between empire, civility, and firmly delineated boundaries could not be more clearly expressed. Britannia, a symbol of British imperialism, is commonly noted for her militaristic garb and masculine Greco-Roman appearance. Britannia demonstrates the ambivalence in expressing the tenets of imperialism in anything but masculine terms. In this particular advertisement, Britannia is the perfect figure for expressing the confusion of gender boundaries that occurs when women enter the male realm of "panoptical" global control. Britannia, in taking up the dust rag, domesticates global time and space while also preserving its masculine image. In 1884, the "concealed alliance" of women's domesticity would make a dramatic reappearance—almost nostalgically—when women's suffrage was overwhelmingly defeated in the House of Commons.

6. Britannia was used to promote the superiority of British products around the globe in numerous late-Victorian advertisements. This ad for silver polish (circa 1895) turns the world into a spotless British home. Photograph from the Robert Opie Collection.

But by 1894, rallies were again being held in London and Manchester in support of adding a women's suffrage amendment to the Registration Bill. David Rubinstein explains that "by early 1894 the demands of women seeking emancipation had reached such a pitch that it was fairly common to write in terms of a 'new' womanhood."[20] The new-woman novel became immensely popular during this time. Sarah Grand's

Heavenly Twins (1893) sold thirty-five thousand copies in its first year and George Egerton's *Keynotes* (1893) was in its seventh edition by 1896. By 1895, Edmund Gosse wrote that "things have come to a pretty pass when the combined prestige of the best poets, historians, critics and philosophers of the country does not weigh in the balance against a single novel by the New Woman." The new-woman novels, which focused on a woman's attempt to overcome social and economic barriers and gain control over her life, generally ended badly. Abigail Ruth Cunningham commented that emancipation, in these novels, tended to bring upon oneself "a relentless catalogue of catastrophe." "Many of the New Women novelists," she continued, "seem positively to wallow in gloom." The consistency of these gloomy endings may point to the fact that the novelists themselves could not imagine a way out of the firmly entrenched boundaries of domesticity—attempts to cross them inevitably led to nervous disorders, disease, and death. It could also be said, however, that the failure of the female characters to achieve independence may have functioned not to reflect a more general social failure but to quell fears of the very real emerging possibility of the disassembling of the Victorian household. By projecting failure onto women who attempted to leave this space, these novels may have reinforced the necessity of remaining within it. New-woman novelist Louise de la Ramée (known as Ouida) wrote that all of these women possessed a "sleeping potentiality for crime, a curious possibility of fiendish evil." This view of women, which was often promoted by new-woman novelists, was also picked up by conservatives reacting to the idea of the new woman. Journalist Hulda Friederichs wrote that the new woman had been "slandered, ridiculed, calumniated, scorned, mocked, caricatured, and abused, till you can hurl no more insulting epithet at any girl or woman than to call her a New Woman."[21]

In contrast to the failed and ridiculed new woman, the suffragettes directly challenged the masochism of these failures and the derision of conservatives by advocating violence in the face of ridicule. In 1897, the respectable *Englishwomen's Review* directly stated that it opposed militant action such as smashing windows.[22] But in 1903, Christabel Pankhurst formed the Women's Social and Political Union (WSPU), an organization that advocated precisely this behavior. In reference to women's window-smashing campaigns, Emmeline Pankhurst said, "We honour these women because having learnt that the argument of the broken pane of glass is the most valuable argument in modern politics, they nerved themselves to use that argument."[23] If the new woman was derided with an almost paternalistic patience, the women of the WSPU were immediately, vehemently, and violently opposed. WSPU member Millicent Fawcett explained: "Instead of the withering contempt of silence, the Anti-Suffrage papers came out day after day with columns of hysterical verbiage directed against our movement. . . . Wherever one went nothing else was talked of; intense hatred and contempt being frequently expressed."[24]

Most historians suggest a distinct shift in women's politics after the emergence of the WSPU.[25] References to window smashing as early as 1897, however, tend to blur that solid dividing line of 1903. And even before women began smashing windows, more general threats of anarchy and crowd behavior were often linked to women's desire to leave the home. Anne McClintock writes: "Images of female violence suffuse the image of the crowd, despite the fact that the unruly urban crowds were predominantly male. Male crowd behavior, it was held, mimicked social behavior that was typical of women." Hysteria, irrationality, docility, fickleness, and gullibility were repeatedly attributed to both women and crowds. Tarde, for instance, wrote that "the crowd . . . is feminine, even when it is composed, as is usually the case, of males."[26] The true resemblance between the crowd and the new woman was that they both appeared to represent a threat to the security of clearly delineated boundaries between public and private; for this reason, they were construed as hysterical and irrational.

Fiery and Polyglot Discourses

Joseph Conrad's *Secret Agent* (1907) captured the public's fear of anarchy in the historical re-creation of the bombing of the Greenwich Observatory in 1894. Interestingly, Conrad's tale of anarchy has been customarily called a domestic drama, for it centers around the relationship between Winnie Verloc and her agent provocateur husband, who convinced Winnie's disabled brother, Stevie, to bomb the observatory. *The Secret Agent,* in many ways, follows the structure of the new-woman novels. Winnie Verloc murders her husband in an attempt to emancipate herself from him, only to lose her mind and kill herself by jumping off a ship to America. Conspicuously located, historically speaking, between the new-woman movement, the Greenwich bombing of 1894, and the suffragette disturbances of 1905, *The Secret Agent* demands further investigation for its analysis of gender, anarchy, and globalization. Surprisingly, Conrad writes nothing about the suffragette movement, even though his book was written at a time when Millicent Fawcett claims "the whole world . . . rang with the doings of the Suffragettes." Nor does Conrad's preface, which was added in 1920, address the militancy and self-proclaimed anarchy of the suffragette movement. *The Secret Agent* appears, instead, to return women to their failed romance with emancipation in the 1890s. Conrad's interest in anarchy, the loss of time, and the murderous impulses of women betrays an anxiety about the relation between gender and globalization that is worth exploring in more depth.

    The Secret Agent is based upon the detonation of a bomb by Martial Bourdin at the Royal Greenwich Observatory on February 15, 1894. The bomb was accidentally triggered by Bourdin en route to the observatory, causing his violent death. W. J. Thackeray, an eyewitness, reported: "The sight that presented itself was pitiable in the extreme, a wretched looking specimen of humanity with his left hand clean blown off just above the wrist, and the sinews and tendons hanging down like bits of cord . . . and a careful examination of the ground around led to discovery of numerous pieces of flesh and bone over the surrounding area." Bone was protruding through "a jagged hole in the back of his coat," and when the doctor opened his vest, he found "his bowels blown in."[27] The identity of Bourdin was quickly discovered because he was found to be carrying a card with his name on it for an anarchist association in town. On the basis of this card, the Autonomie Club was stormed, and the press reported on February 16, "The Metropolitan Police have discovered what is believed will prove to be the most desperate and dangerous of any revolutionary plot that has ever had its headquarters in London."[28]

    The gruesome details of Bourdin's death were quickly released, which drew hordes of tourists to the scene of his death, where they could view the white stakes in the grass that marked the sites where "pieces of his hand and other particles" had been found. Even a steady rain on Sunday, one paper reported, could not keep away the "hardy and determined sightseers," who came in "constant succession." A reporter wrote: "One of the keepers stated that in all his experience he had never known so many people visit the park on a single day . . . even when the band was playing in the summer months."[29] The story headlined every paper for days afterward, a fence had to be constructed to hold the crowds back from the crime scene, and a general and long-lived discussion of the role of anarchists in London began on that day. On February 16, "Bombs and Anarchy," an "Extra Special Edition" of the *Evening News and Post,* came out. Though no one, at that point, was sure why the explosion had occurred, the news quickly spread of the man's identity and associations. The *Post* explained: "By some accident his explosive sent him to eternity—a fact for which society should be profoundly grateful. Terrible as his death was we cannot pretend to feel the least pity for him."[30] The *Illustrated London News* concurred:

> An accidental death, hideous and horrible, but scarcely deplorable, as it deservedly ended the pernicious existence of one of those detestable criminals who plot the wholesale murder of the innocent, the destruction of private and public property, and every other cruel mischief that fiendish cunning devises for the vain purpose of terrifying society to overthrow all social and political institutions, took place on Thursday afternoon, Feb. 15, in Greenwich Park.[31]

The *Illustrated London News* concluded: "A Frenchman named Martial Bourdin, has, fortunately for mankind, killed himself unintentionally with the vile and dreadful instrument that is in vogue among them for their absurd and monstrous schemes."[32]

The day after the Greenwich explosion, a list of suspected anarchists in London was printed in the *Morning*. The *Morning* justified this list by stating: "There can be no doubt that at the present moment the West-end of London is swarming with the friends and companions of Vaillant and Henry [Bourdin], who would be only too pleased if they could blow up a building. . . . They may be Frenchmen, Spaniards, Russians, or Englishmen, but, first and foremost, they are the foes of civilization."[33] Anarchists of all nationalities, it was suddenly announced, had been making "free and unmolested use of the metropolis" for some time— the Autonomie Club, frequented by the brothers Henry and Martial Bourdin, had apparently been "known to the police for years past as being frequented by political desperadoes of all nationalities, wherein Anarchy and the 'Social Revolution' is preached and advocated nightly in fiery and polyglot discourses."[34] But the club was allowed to operate to keep the "Anarchists under surveillance" until the Greenwich bombing. After February 16, anarchists were repeatedly described as monsters or lunatics whose "polyglot discourses" alone seemed to supply proof of their incomprehensibility. The *Evening News and Post* announced that England must "join hands with our Continental neighbours in exterminating the Anarchist breed."[35] And another article suggested: "That Anarchists as a matter of fact . . . are lunatics is as obvious to commonsense as that a dog is mad when he rushes at friends and foes indiscriminately."[36] Anarchists became constructed as a mad and animalistic race whose very multilingual capacities made them untrustworthy. Conrad, in *The Secret Agent,* is also notoriously unsympathetic to anarchism. He reflects, in the introduction, on "the criminal futility of the whole thing, doctrine, action, mentality." Conrad writes that anarchy is bent upon "exploiting the . . . passionate credulities of a mankind always so tragically eager for self-destruction"—for this reason, he explains, it is "unpardonable."[37]

THE MATERNAL FLAME

Conrad's idea for *The Secret Agent* came about when he was reading the memoirs of the assistant commissioner of police at the time of the bombing. He noted his thoughts as he read:

> One fell to musing . . . of South America, a continent of crude sunshine and brutal revolutions, of the sea, the vast expanse of salt waters, the mirror of heaven's frowns and smiles, the reflector of the world's light. Then the vision of an enormous town presented itself, of a monstrous town more populous than some continents and in its manmade might as if indifferent to heaven's frowns and smiles; a cruel devourer of the world's light. There was room there to place any story, depth enough for any passion, variety enough there for any setting, darkness enough to bury five million lives. . . . Endless vistas opened before me in various directions. It would take years to find the right way! It seemed to take years! . . . Slowly the dawning conviction of Mrs. Verloc's maternal passion grew up to a flame. (40–41)

If one were to analyze the trajectory of Conrad's thoughts, it appears that anarchy put him in mind of revolution, which made him think South

America. The "light" of South America, however, became eclipsed by the darkness of an "enormous town," presumably London. What inspired Conrad's story, then, is not the bombing itself, but the "monstrous town" of "man-made" London, which, in contrast to South America is a "cruel devourer of the world's light" in which one necessarily becomes lost.

Turning working-class London into a place of darkness was a common literary convention in the 1880s and 1890s. In 1890, William Booth published In *Darkest England and the Way Out,* which was directly inspired by Henry Morton Stanley's *In Darkest Africa.* Conrad, like many other middle-class young men at this time, commonly ventured out as an urban explorer of darkest London. *The Secret Agent* represents Conrad's attempt to negotiate the urban slums as "epistemological problems—as anachronistic worlds of deprivation and unreality, zones without language, history or reason that could be described only by negative analogy."[38] He explained, "I had to fight hard to keep at arm's length the memories of my solitary and nocturnal walks all over London in my early days, lest they should rush in and overwhelm each page of the story" (41). Conrad appears to be preoccupied with evoking that threshold of darkness without being overcome by it and letting it "rush in and overwhelm." The streets of London, in *The Secret Agent,* are continually described as "sullen, brooding and sinister" (152), and Conrad stages the novel as a kind of battle against these memories of dark London.

But it is Mrs. Verloc who lights the novel for Conrad; it is her "flame" that is the light guiding us through the streets of London. Mrs. Verloc's "maternal passion" somehow keeps the city out, creating a safe haven for her husband and brother while lighting a path for the reader. *The Secret Agent* is a domestic drama, as Conrad himself insisted and the assistant commissioner points out when investigating the explosion: "From a certain point of view we are here in the presence of a domestic drama" (204). The narrative turns into a domestic drama precisely because of the contrast between being lost in an overwhelming city and finding a place called home. The house is described by the narrator: "Ensconced cosily behind the shop of doubtful wares, with the mysteriously dim window, and its door suspiciously ajar in the obscure and narrow street, it was in all essentials of domestic propriety and domestic comfort a respectable home" (185). It is the house, and specifically the bed, that Stevie sees as the solution to London's problems with crowding and poverty; in his desire to "take the whole world to bed," Stevie suggests the alternative to bombs and explosions (166). Though Verloc's home also functions as a pornography shop, it is ironically the closest thing to domestic security in the novel.

Adolf Verloc's notion of domesticity is accompanied by a sense of his possessions. He is said to have loved his wife "maritally, with the regard one has for one's chief possession" (174). She provides him with a sense of "familiar sacredness—the sacredness of domestic peace" (220). Winnie, on the other hand, marries Verloc only because of her brother Stevie, who is mentally disabled and therefore—she believes—in need of a kind benefactor. She calls her marriage "seven years' security for Stevie," which is "loyally paid for on her part"—this security then grew "into a domestic feeling" for her. Yet when the "payment" of herself is no longer needed because of the death of Stevie, she experiences an overwhelming sense of freedom: "There was no need for her now to stay there, in that kitchen, in that house, with that man—since the boy was gone for ever. No need whatever. And on that Mrs Verloc rose as if raised by a spring. . . . Mrs Verloc began to look upon herself as released from all earthly ties. She had her freedom. . . . She was a free woman" (226). Christopher GoGwilt remarked that "with a deliberate antifeminist irony," Conrad repeatedly refers to Winnie as a "free woman" after Verloc's death.[39]

Winnie kills Verloc in her newly inspired freedom, as an act of revenge for Stevie's death. But she is then completely befuddled as to how to use her new freedom: "She had thrown open the window of the bedroom either with the intention of screaming Murder! Help! or of throwing herself out. For she did not know exactly what use to make of her freedom. . . . The street, silent and deserted from end to end, repelled her by taking sides with that man who was so certain of his impunity" (228). When Winnie enters London, lost and alone, her cognitive-mapping ability fails, and we can see that a distinct gender difference emerges in cognitive-mapping skills. Adolf may feel that the city is a "maze," but he still feels that he is "cosmopolitan enough" to outwit its labyrinthine traps. But to Winnie, the city of London is a blank, as the narrator comments: "The vast world created for the glory of man was only a vast blank to Mrs Verloc. She did not know which way to turn. . . .

She was alone in London: and the whole town of marvels and mud, with its maze of streets and its mass of lights, was sunk in a hopeless night, rested at the bottom of a black abyss from which no unaided woman could hope to scramble out" (240). When she drives across London, the narrator writes: "Night, the early dirty night, the sinister, noisy, hopeless, and rowdy night of South London, had overtaken her on her last cab drive" (159).

Because she cannot pull herself out of this city "unaided," Winnie seeks the aid of Ossipon, who had pursued her with "shamelessly inviting eyes" since they first met (220). "Take me out of the country," she begs him. "I'll work for you. I'll slave for you. I'll love you. I've no one in the world. . . . Who would look after me if you don't!" (253). But Ossipon, who at first agrees, loses faith in Winnie after he learns of her dead husband. Suddenly, she is incontestably marked in his eyes, becoming a kind of Medusa figure with snakes emerging from her hair. Earlier, Verloc had made similar associations when he looked at his wife; he "gazed at her, and invoked Lombroso" (259). Cesare Lombroso (1835–1909) was an Italian criminologist who attempted to prove that certain facial characteristics indicated criminal psychopathology; he once made the infamous claim that "even the normal woman is a half-criminaloid being."[40] Conrad, in invoking Lombroso, plays into stereotypes of women that had also been repeatedly invoked by the new-woman novelists. Ossipon, in seeing Winnie's "half-criminaloid being," rejects her and leaves with her money. Winnie, desolate and alone, jumps from the boat on her Atlantic crossing to America.

Conrad claimed it "would take years to find the right way" in the city of London. Winnie, ironically, functions as a guide in the tale, for she fills the same idealized role that Conrad sees in South America. Her passion is a "light" through the darkness of London, just as Conrad describes the sea between South America and Britain as a "reflector of the world's light"—and South America as a land of "sunshine." Winnie appropriately dies in that same sea on her way to America. The light is cast out of the city, which continues in its darkness and security. It is free from the impositions of suffragettes, who cannot burn it down, and free from anarchists, who cannot explode its strength. When Martial Bourdin died, one newspaper commented, "By his death, it is thought, the whole fabric of a great Anarchist plot has been destroyed."[41] If Conrad was creating an analogy between domestic explosions and anarchistic ones, it would appear that Winnie posed a threat to the security of property as much as to Verloc. But the threat proves futile, and Winnie fails, cast off to the eternal light of South America.

LONDON, WITHOUT TIME

The Secret Agent was published in 1907, the same year that Finland became the first country in Europe to give women the vote. Though Conrad hearkens back to the time of the new-woman novel and the Greenwich bombing, his readers easily may have been contemplating what the vote for women would mean. By combining a tale of feminism and anarchy, he also makes associations that the reader could easily conflate. The feminists were the anarchists in 1907, as Emmeline Pankhurst herself pointed out: "People are saying . . . that women are creating anarchy. No, it is not we who create anarchy. Anarchy is there. There is anarchy in a country which professes to be under constitutional and representative government, and denies the benefits of the constitution to more than half its people."[42] Pankhurst turned the stereotype of women's criminality on its head by embracing, rather than evading, the label of criminal. Pankhurst claimed, "I do not look either like a soldier or very like a convict, and yet I am both." Women, she continually implied, are not what they seem; they are more—not less—dangerous than you thought. Continually fighting notions of women's docility and passivity, Pankhurst believed, like Frantz Fanon, that violence was key to reshaping the identity of an oppressed group. She attempted to demonstrate, through her acts of guerrilla warfare, that women—just like any other group—could be revolutionaries. "If an Irish revolutionary had addressed this meeting," Pankhurst said at an address at Hartford, "it would not be necessary for that revolutionary to explain the need of revolution beyond saying that the people of his country were denied—and by people, meaning men—were denied the right of self-government. . . . But since I am a woman it is necessary in the twentieth century to explain why women have adopted revolutionary methods in order to win the rights of citizenship."[43] The revolutionary methods that Pankhurst advocated included breaking most of the windows on Regent and Oxford Streets, placing a homemade time bomb in the home of Lord George, and burning down railway stations.

Emmeline Pankhurst clearly stated the methodology behind these attacks:

> We have set ourselves to force this Government out of office, and I know that business men in this country will help us. Your stockbroker, whose communication with Glasgow was cut off for several hours during very important business hours, does not want that to be a weekly occurrence. Your business man who reads important communications through the post does not want his communication to be interfered with as a regular and permanent institution. . . . Your business man, whose customers mostly are women, does not like the idea that in the interval of buying hats, his customers may be breaking his shop windows.[44]

The enemy, to the suffragettes, was property, as Emmeline Pankhurst had earlier explained: "There is something that the Government cares for far more than they care for human life, and that is the security of property. Property to them is far dearer and tenderer than is human life, and so it is through property we shall strike the enemy . . . it is only as an instrument of warfare in this revolution of ours that we make attacks on property."[45] In attacking property, communication systems, and men's clubs, suffragettes attempted not only to irritate and gain publicity, but also to disrupt the complex structure of industrial rationality. Brian Harrison explains, "Mrs. Pankhurst's followers provided a standing demonstration of the guerrilla's nuisance value to governments enmeshed in the complexities of modern industrial society."[46]

Because communications were regularly targets of attack, Scotland Yard, which had created a special department to deal with the suffragettes, became concerned that the Greenwich Observatory would be targeted. In 1915, F. W. Dyson wrote, "The Scotland Yard Authorities, at the time of the suffragette disturbances, thought it necessary to have police guarding the observatory." A letter addressed to the Royal Observatory had warned: "A gentleman called at the East Ham Police Station on Sunday night and reported that he heard two well-known suffragettes conversing on a tram car. He heard them say, 'Wait till they start on the Greenwich Observatory; London, without time, will cause them to wake up.' In view of this report the Superintendent advised police protection of the Royal Observatory."[47]

The Royal Observatory, by the turn of the century, had come to signify the security of standardized time and space. It kept the trains from crashing, allowed communication to distant locations by telegraph, organized shipping schedules, and generally came to represent the benefits of globalization. In *The Secret Agent,* when Vladimir and Verloc are trying to decide on the best target to blow up, Vladimir says: "Go for the first meridian. You don't know the middle classes as well as I do. Their sensibilities are jaded. The first meridian. Nothing better" (70). The infrastructure of the economy is based upon the correctness of time; thrown back upon local time, trains would not reach their destinations, ships would be lost at sea, and trade itself would be wrecked. Vladimir intimates the significance of the meridian to Verloc: "Since bombs are your means of expression, it would be really telling if one could throw a bomb into pure mathematics. But that is impossible. . . . The blowing up of the first meridian is bound to raise a howl of execration" (67–68). Vladimir claims that science had replaced religion, art, and even education as the icon of the public. The Greenwich Observatory stands for the scientific ordering of space, which preserves England as the center of empire.

Women disrupted the sacred space of men's clubs, like the Royal Geographical Society, and posed a threat to the globalized space of industrial rationality. In this sense, theirs was an agenda of occupation and disruption. In response, the antisuffrage movement vehemently declared that the women's place was in the home. In 1913, the *Oxford Times* reported that "a fetid mob (I can use no other words)" threw rotten eggs at the suffragist leaders, while singing, "There's no place like home" and "Go home and mind the baby."[48] Women, violently evading these demands, threatened not only to occupy male-defined space, but also to demolish it. The suffragettes disrupted the space of Greenwich without ever bombing the observatory itself; their program of attacking men's clubs and communication centers threatened a government that was increasingly dependent on standard time.

Greenwich promised to organize time and space around the premise of objective rationality. But as Frantz Fanon has suggested, "For the native, objectivity is always directed against him."[49] Similarly, the emergence of objective space served to dispossess women by forcing them back into the home. The delegates to the International Meridian Conference, all male, set the standard for time and space, while the demand for women's suffrage was ignored in England. In *The Secret Agent,* Conrad represents the threat to time and space by resurrecting the body of Martial Bourdin. At a time of increasing civil unrest from women, the Irish, and the colonized around the world, Conrad highlighted both a failed attempt at anarchy and a woman's failure to "emancipate" herself from the city of London or her husband. The threat of "London without time" would be raised, and eliminated, in this novel; the attempt at "anarchy" proved to be merely self-destructive. The suffragettes of Britain would fare better, ultimately gaining the vote, and thus officially entering the public sphere, in 1918. Ironically, as women began to take part in the British political system, their identification with the Irish and other colonized groups was largely forgotten. Women were suddenly no longer extranationals, insisting on their ungovernability and threatening to undo the government itself. They became, very quickly, accomplices rather than dissidents, imperial-ists rather than anarchists. They accepted Greenwich as their own, though the struggle of the colonized against English rule would continue after—and often be modeled upon—the victories of women.

This historical moment—when the war against domestic space par-alleled the war against global space—allowed women like Christabel Pankhurst to redefine their gender and their place in the imperial proj-ect. Elizabeth Grosz has written, regarding the role of feminist geog-raphies: "The project ahead is to return women to those places from which they have been dis- or re-placed or expelled, to occupy those positions—particularly those which are not acknowledged as positions—partly in order to show men's invasion and occupancy of the whole of space as their own and thus the constriction of spaces available to women."[50] Indeed, according to Grosz, changing representations of sub-jectivity are associated historically with changes in representations of time and space. The women's movement, in its beginnings, was also a struggle for women to gain a space in universal time. Crossing the boundaries of time and space, in an effort first to destabilize and then to remake it, women crossed the boundaries of gender itself. If "the project ahead is to return women to those places from which they have been dis- or re-placed or expelled," one of those places would be time. But rather than return women to their place in universal time, I have instead attempted to remember their moment extratemporally, or in-between times, which threatened to undo time.

Today, Martial Bourdin's body has become lost in the archives of the Greenwich Royal Observatory. The time of the "railway clerks" has been largely naturalized, and the "desperate attempts to 'annihilate both Space and Time'" that the astronomer royal once lamented have succeeded. I visited the Royal Observatory in 1998, looking for the one photo of Martial Bourdin that I knew still existed, only to be told, "I heard that so-and-so in inventory saw it once." Upon my calling that person, the rumor was denied and my motives for wanting to see it were questioned. Other employees began to talk about why the explosion might have occurred: "I heard he was trying to protest something, but I'm not sure what," one person said. "He was protesting the crowding at the park," another joined in. The official policy of the observatory, today, is to suppress this photo, claiming it is "too sensitive" to release to the public. Still, the photo carries all the intrigue of a suppressed rumor among the observatory staff; some have heard of its existence or even seen it, but even they will only whisper this experience to those who refuse to believe it is true. Bourdin's body still circulates as the suppressed, mythological undercurrent to the certainty of time and space. The prime meridian promised to stifle multiple notions of temporality, to homogenize space, and to provide a means for communicating global coordinates. It opened up space as a global, translatable phenomenon—and in this sense, rendered it safe and "universally" understood. It minimized paperwork. But, though the death of Bourdin may have symbolically killed "the whole fabric of a great Anarchist plot," it still circulates as the suppressed historical narrative about the struggle over time. Similarly, the suffragette promise of "London without time" serves to problematize the place of women in the continued production of global space, the nation, and the history of empire.

CHAPTER 2

Pundit A and the
Trans-Himalayan Surveys

AT A MEETING of the Royal Geographical Society in 1899, Thomas Holdich announced the opening at Oxford of England's first school of geography. He claimed that this school would provide aid for "recent frontier operations" in Asia and would include "representatives both of the navy and the army." Holdich, an explorer of Tibet, went on to promote the supposed impartiality of triangulation, suggesting that this would be the obvious focus for the school: "Every point on a boundary-line, every peak in a mountain system . . . has a value whose correctness can be proved just as easily in a London office as in the open field. And this value is not only incontrovertible, but absolutely distinctive, because every point on the whole world's surface has its own special position in terms of latitude and longitude, with which no other point can interfere." This melding of military discourse and the language of incontrovertibility would serve to promote England's role in the colonies as impartial arbiter in territorial disagreements. Holdich continued: "Just as the Providence of battles usually favors the biggest battalions, so it is likely that the widest geographical knowledge will prove the best safeguard against misunderstandings."[1] It is no coincidence that this comparison was made between big battalions and big maps; in the "frontier operations" of Asia, England would serve as both definer and implementer of boundaries—defining through cartography and implementing with battalions.

Triangulation surveys were linked with military control from their inception during the Highland Uprising of 1745 in Scotland.[2] Showell Styles remarked, in *The Forbidden Frontiers,* that surveyors commonly "went ahead to make the rough but trustworthy maps by which the campaigns were fought."[3] In India, the Great Trigonometrical Survey, which began in 1818, was labeled one of "the most glorious monuments of British rule in the east."[4] By 1899, the triangulation of India was complete, and the focus of geographers had shifted to fixing boundaries with China in the north and Russia in the northwest (fig. 7). Russia had been triangulating southward as England was triangulating northward from India; those countries that were caught in the middle—Turkestan and Afghanistan—became areas of contention. Henry Rawlinson of the Royal Geographical Society commented, "It was in reality the debatable ground between India and Russia and must naturally become more interesting year by year as we went on towards our future destiny."[5] Penetrating these zones became the fixation of geographers at the end of the nineteenth century—and the language of destiny, protection against enemies, and the necessity of science became the rhetoric that justified this desire.

Tibet resisted rule by either Britain or China, and Afghanistan's independent Afridi were adamantly opposed to invasion. According to

7. India, covered in triangles, was the geometric ideal of the Great Trigonometrical Survey. "Index chart to the Great Trigonometrical Survey of India, completed under the orders of Col. J. T. Walker, 1876," photograph from the Royal Geographical Society, London.

Styles, "Nothing would ever persuade the natives of the independent states that a survey of their territories was not the preliminary to an invasion."[6] These in-between areas, in turn, created an unreliable third space in colonial mapping projects. The residents of these contested territories were often seen as problematically interfering in the rightful dialogue between empires. According to L.J.L. Dundas, when a neutral zone is "peopled by semi-barbarous tribes" and a "strong and highly civilized power" is nearby, the "civilized power" will always "gradually push forward to the real frontier" until the neutral zone is "absorbed."[7] This notion of the "frontier" as "proximity to the cultural zero" is continually repeated throughout colonial discourse.[8] Real boundaries, the implication seemed to be, could only be between two "civilized" powers.

It is this slippery third space within mapping that I would like to examine. For in reality, this space functions not only as a blank space or frontier on the map, but also as a metaphor for the native mind. Styles explains this phenomenon: "In a sense, the altruistic rulers of British India before the Empire were striving—like the men of the Survey—towards a forbidden frontier, the frontier where the mind of Indian and Englishmen would be as one mind."[9] In identifying this space as a neutral zone, an attempt is made to neutralize racial anxieties about independent, and therefore uncontrollable, indigenous peoples. If the gap could be closed between India and Russia, it was believed, the gap between the native and English mind would close as well, by eliminating other points of interference. David Spurr has suggested that, within colonial discourse, "the desire to emphasize racial and cultural difference as a means of establishing superiority takes place alongside the desire to efface difference and to gather the colonized into the fold of an all-embracing civilization."[10] The Indian northwest frontier represented both these things. The purpose of marking the frontier was to establish an ordered hierarchy between races in the face of conjectured disarray; at the same time, it was believed that closing the frontier would mean absorbing these races into the "one mind" of England.

Crossing the frontier into resistant territories necessitated entering this "vast terra incognita" incognito. Europeans often had to rely upon native spies, leading to a form of espionage that required racial hybridity or camouflage. Thus, the neutralization of race fails, and negotiating the problem of hybridity takes over cartographic discourses. Initially,

Europeans attempted to dress as Indians in order to enter Tibet; when that proved too dangerous, they sent Indians instead. It is at this point of mapping in disguise that "monstrous hybridities" would begin to emerge, a term that Rudyard Kipling used to describe Hurree Babu in *Kim*, a character based upon the real explorer Sarat Chanda Das. Hybridity, as a sign of European exclusion, would begin to be seen as monstrous. That is to say, the investment in fixing territorial boundaries as neutral overrides the investment in strict racial categorization. The secure whiteness of the map required the allowance of temporal hybridities, which are nonetheless defined as "monstrous." The disidentification of the map with race, in this sense, allowed for an often ambivalent racial play to emerge, which is nowhere more evident that in Rudyard Kipling's *Kim*.

The Pundit Expeditions

Kim is based upon the pundit expeditions of the 1860s, in which native Hindus called "pundits" traveled in disguise into Tibet for the purpose of surveying. T. G. Montgomerie devised this method in 1863 because Tibetans would not allow Europeans—particularly surveyors—into their country.[11] The pundits were trained by Montgomerie to take bearings with a compass, to pace the distances, and to determine latitude with a sextant. The results of these observations, however, could only be determined by their British superintendents at the Trigonometrical Survey office, in marked contrast to the standard British practice of computing results in the field. Clements Markham explained the reason for the difference between British and Indian methods:

> They [the pundits] are purposely not taught how to reduce their observations, nor supplied with astronomical tables, in order that they may not be able to fabricate fictitious work. Observations for determining absolute or differential longitudes are beyond their capacities. The resulting latitudes and the co-ordinates of the results are computed in the office of the Superintendent of the Great Trigonometrical Survey on the explorer's return.[12]

This fear of Indian fictionalization of the map demonstrates that the pundits, in part, were viewed as potential threats to the very maps they were creating. The belittling of their abilities, in turn, serves to verify

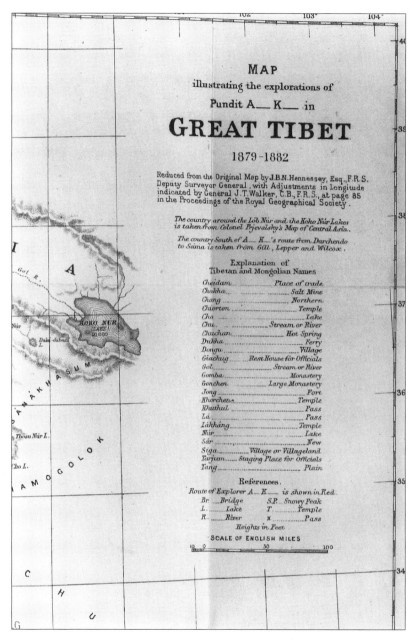

8. Maps like this depicted the routes of the pundit mapping expeditions, but the pundits were forbidden to draw maps themselves. "Map Illustrating the Explorations of Pundit A__ K__ in Great Tibet, 1879–1882," photograph from the Royal Geographical Society, London.

the abilities of the Europeans. In both instances, these excuses may have been used simply to deflect credit to the European as the author of maps (fig. 8).

The other major difference between British and native explorers was that the pundits were addressed by a number or letter, rather than a name. In 1864, Montgomerie employed "Pundits A and B" to go to Lhasa.[13] Pundit A, or Nain Singh, was also more commonly called "Number 1" or simply "the Pundit." Singh's cousin Mani Singh was called "G-M," and another relative, Kishen Singh, was called "A_K_" (fig. 9). In 1866 and 1867 another expedition was made "by an explorer whom Colonel Montgomerie calls No. 9." And in 1872, Montgomerie organized a party "led by a young semi-Tibetan, who is neither distinguished by name or number, so we will call him D."[14] Overall, a dozen or so Indian explorers were employed between 1864 and 1884, most all

9. Pundit Kishen Singh, who traveled in the company of a lama into Tibet in 1881. Photograph from the Royal Geographical Society, London.

of them assigned numbers or letters. Although Nain Singh would become the most famous, for determining the correct position of Lhasa, other exploits would gain notoriety, such as Kinthup's story of escaping slavery after two years to complete his task of throwing five hundred logs into the Tsangpo River to chart its course. He wrote ahead to explain that "the journey proved a bad one" but that "I, Kinthup, have prepared 500 logs according to the order of Captain Harman, and am prepared to throw fifty logs per day into the Tsangpo."[15] His letter, unfortunately, was not received in time, and the logs came down the Tsangpo unseen.

In 1864, "Pundit A" set out, disguised as a native from Ladak and a companion of a Tibetan merchant.[16] This pundit, Nain Singh, was a Bhutiya subject who, at the suggestion of Montgomerie, carried a Buddhist rosary and a prayer wheel. Rather than carry prayers inside the prayer wheel, however, Singh carried slips of paper to record his compass bearings. And the rosary, "which ought to have 108 beads, was made of 100 beads, every tenth bead being much larger than the others . . . at every hundredth pace a bead was dropped."[17] Singh returned not only with a journal of his impressions of Lhasa, but also with enough information to create a map that would be the first geographical record of the country. Singh's journals, interestingly, are quite critical of life in Lhasa. He writes, "I observed that there was but little order and justice to be seen in Lhásá"—and he describes scenes of superstition, usury, and archaic lifestyles (161). For instance, he complains of the election process in Lhasa, suggesting that the judge "exercises his authority in the most arbitrary manner possible, for his own benefit, as all fines, &c., are his by the purchase" (163). Much of his journal also appears to be devoted to recording the types of trade that occur in Tibet, as well as the types of merchandise that could potentially be traded (silk, wool, tea, rice, etc.).

Singh's journal, while critical, is quite consistent with the European demands for information regarding trade and local government, and so it may provide more evidence about the demands of the audience than the opinion of the author. At a meeting of the Royal Geographical Society, Henry Rawlinson described these expeditions: "Geographical discovery led to the spread of civilisation and general intelligence, and even to material advantage in the advancement of commerce and trade. . . . The Pundit's travels in Tibet had paved the way for the extension of our trade in that direction." If Tibet and the neighboring countries were thoroughly explored, he continued, "tea might penetrate from India, if not from China, into Turkestan, by a hundred different channels." The exploration of Tibet would also, he claimed, "prove of very great value in improving the social state of the East" (170). The penetration of tea into Tibet, therefore, became equated with the improvement of social conditions—and geography was the means to this end.

Nain Singh's journal was presented at the Royal Geographical Society by Captain Montgomerie; Singh was interestingly absent, though Montgomerie commented that seeing his face would have made his journal more believable. The presentation of the journal was followed by an elaborate discussion about the veracity of natives, along with a celebration of Montgomerie's success. The discourse about Singh is almost as interesting as—and much more voluminous than—his own journal. The president of the Royal Geographical Society praised Montgomerie, "without whose admirable and ingenious contrivance of instructing an intelligent native, and sending him in disguise, the Society would never have had this account of the country brought before them" (165). These expeditions were characterized as "native enterprise directed by English intelligence" (169). And although the pundits put themselves in grave danger for these journeys, Montgomerie's heroism grew with every trip. In 1868, the president stated:

> Captain Montgomerie had every reason to be proud. These Pundits had been trained to penetrate these difficult countries, acquiring the languages, and being instructed how to make observations, which rendered their journeys of high scientific value. . . . They travelled at risk of their lives every moment; for if one of the scientific instruments which they possessed had been detected in their boxes, they would have been put to death. Animated by *esprit de corps,* and a love for science, these Pundits had been able to traverse the country where no European would be safe, and to make a series of observations for latitude and longitude.

This passage reveals the way in which praise was easily and continually deflected from the pundits to Montgomerie. Although the pundits had traveled "at risk of their lives every moment," it was Montgomerie who had "reason to be proud." What, then, could have motivated Nain Singh to perform this work at all? According to the president, "He had much pleasure in making these observations, because he saw in front of him two distinguished Indian authorities." These Indian authorities were Robert Montgomerie and the ex-governor-general of India, Lord Lawrence, who had approved the arrangements made by Captain Montgomerie.[18] According to the president, Singh was led forward by the faces of these government officials, even though they were not physically on the journey. Again, any motivation on the part of Singh is directed back to the great honor of white men. The president writes, "These native explorers did a good service in the field, but they were quite unable to utilize their work, and for the resulting narratives and maps geographers are indebted to . . . Colonel Montgomerie."[19] Even the narrative component to these journeys, it seems, is credited to the white leaders of the expedition.

Like the "semi-barbarous" people of the neutral zones, the narratives of the pundits are absorbed into a larger European discourse of latitude and longitude. Interestingly, while the pundits are said to be inspired and drawn forward by the faces of the "two distinguished Indian authorities" before them, they make the observations only "for latitude and longitude." Latitude and longitude alone are the indisputable leaders, who easily displace even the faces of white men. The doubly displaced subjectivity of the cartographer-pundit demonstrates the shifting role of race in relation to cartography. By turning the conquest of foreign territories into the "incontrovertible" science of latitude and longitude, territories may be secured merely as points on a map that preclude racial distinction. Indeed, the goal of the cartographer is to eliminate racial identity as much as possible because it may disrupt the language of latitude and longitude—which is ultimately a white language. That is to say, it is considered transparent and universal and governed by Europeans. In Kipling's *Kim,* this racial anxiety and elimination would be represented through Pundit A becoming white—as the character of Kim—deflecting "monstrous hybridism" back onto the East.

WHITE PUNDITS

Published in 1901, Rudyard Kipling's *Kim* recreated the journeys of the pundit explorers. Kim, the boy who takes the role of a pundit, however, is Irish. And although Kipling's characters travel the routes of native explorers of Tibet in the 1860s, the novel becomes a way of answering the problems of India's frontier policy in Afghanistan, an issue that was current at the time of publication. The frontier policy debate centered around a dispute between what was then called the "forward school," which favored invading India's neighboring territories, and the "stationary school," which favored remaining on the Indus plains. In either case, the eyes of the British public were drawn to the frontier in order to engage in the debate.

Kim is a novel about watching the passes—a Russian surveyor has been discovered heading over the Himalayan passes into India, accompanied by a Frenchman. Kim's role in this "great game" is to steal the maps and journals of these explorers, so that information about British territory will not fall into Russian hands. (The term "great game" originated during the first Afghan War in 1839 and was later commonly used to describe British relations with Russia.)[20] To accomplish his task, Kim brings along a Buddhist lama and disguises himself as a follower—also, he is accompanied by Mahbub Ali, a Muslim who aspires to be an ethnologist for the Royal Society. Mahbub Ali's intention, therefore, is to study the lama as Kim is scouting out the route of the Russian.

Raised by a "half-caste" opium dealer who "pretends to keep a furniture shop," Kim has a background as questionable as his identity. His mother and father were Irish, but his mother died of cholera and his father—once in an Irish regiment—took a post on the railroad and became acquainted with a "furniture dealer" who introduced him to opium. Opium then became his addiction until he "died as poor whites die in India," leaving Kim in the hands of the opium dealer. Kim's identity appears to be predisposed to the rules of the great game before he even engages in it, as the narrator remarks:

> Kim's ability to slip in and out of Hindu, Buddhist, Mohammedan, or European identities, changing from one to the other as the whim or need strikes him, is part and parcel of his being a "Friend of All the World" and continuous with his ability to scuttle from housetop to housetop at night, without being limited by the hindrance of being confined within any one particular home. To be any less polymorphous would be to be capable or even desirous of having friends, relatives, and social constraints on one's movements that would prevent one from playing "in" one's own life as if it were a "game."[21]

Kim's nickname, "Friend of All the World," coincides with his ability to turn his own identity into a game, changing disguises and friends as need permits—which ironically makes him incapable of having friends or family.

Kim grows up not understanding his past or ethnicity, though he wears a charm that his father gave him, which carries his birth certificate and other identity papers. After he discovers his parentage, Kim thinks: "'I am a Sahib. . . . No; I am Kim. This is the great world, and I am only Kim. Who is Kim?' He considered his own identity, a thing he had never done before, till his head swam" (126). Interestingly, it is not until his charm is deciphered by a white man that Kim begins to question his identity. It is as if discovering that he is a specific human being in the midst of an overwhelming nation is too much for him. Kim begins a ritual of repeating his name like a mantra at moments of confusion: "I am Kim—Kim—Kim—alone—one person—in the middle of it all" (225). The problem for Kim, however, is not his Irish identity; the problem is consistently defined in terms of how he fits into India after learning that he is white. He thinks to himself: "'Among Sahibs, never forgetting thou art a Sahib; among the folk of Hind, always remembering thou art—' he paused, with a puzzled smile. 'What am I? Mussalman, Hindu, Jain, or Buddhist? That is a hard nut'" (149). The irony of Kim's disorienting identity is that it may, in fact, make him a better recorder of details. The narrator describes Kim as a boy who "by virtue of being nothing itself is all the more capable of seeing and hearing everything around it" (8). In order to have an eye for detail one must, Kipling implies, maintain an objective or scientific perspective. Kim, as a boy lost in the game since birth, is all the more equipped for recording the "true" India.

THE REALITY EFFECT

Literary critics have long praised *Kim* for presenting a realistic picture of India. Abdul JanMohamed claims that *Kim* introduces us to "a positive, detailed and nonstereotypic portrait of the colonized that is unique in colonialist literature." He continues, "What may initially seem like a rapt aesthetic appreciation of Indian cultures turns out, on closer examination, to be a positive acceptance and celebration of difference."[22] *Kim* is in fact filled with stereotypes, criticisms, and fabulous

descriptions of the colonized; but it is that aesthetically fabulous realm
of description that makes the novel appear to celebrate difference.
Patrick Williams comments on the remarkable way in which *Kim,* an
"eminently fabular work," has been consistently "read as the supreme
example of realist fiction in India."[23] The parameters of this debate are
themselves fascinating. Williams suggests that *Kim* creates what Barthes
calls a "reality effect"—and it is precisely this effect that warrants further
discussion, rather than the verity of the text itself.

JanMohamed suggests that *Kim* creates a "portrait of the colonized,"
and it is this language of portraiture, in and of itself, that is important. A
colleague of Kipling in India, E. Kay Robinson, described his work: "It
is this which is the strongest attribute of Kipling's mind: that it photographs, as it were, every detail of passing scenes that can have any future
utility." While Robinson was speaking of the utility of photographs for
Kipling's literary pursuits, there is a way in which his trained photographic eye may be recording India for military pursuits. Kipling,
Robinson wrote, possessed "a marvellous faculty for assimilating local
colour without apparent effort."[24] It is the sheer effortlessness of
Kipling—who, like Kim, can automatically record Indian culture—that
permeates the text of *Kim.* Kim, who instinctually participates in intrigue
as a boy, is drawn to the game for the pure pleasure of stealth: "What he
loved was the game for its own sake—the stealthy prowl through the
dark gullies and lanes, the crawl up a water-pipe . . . and the headlong
flight from housetop to housetop under the cover of hot dark" (21). He
is drawn to the lama with the same simple excitement of a promised
adventure: "This man was entirely new to all his experience, and he
meant to investigate further: precisely as he would have investigated a
new building or a strange festival in Lahore city. The lama was his trove,
and he purposed to take possession" (30). But the language of possession
promises more than the mere satisfaction of curiosity; just as Kipling was
said to be able to assimilate local culture, so Kim's interest in the lama
appears to have more to do with an absorption of his character and the
knowledge he has to offer than a dialogic relationship. In this sense, *Kim*
has a much different role than merely presenting a "celebration of difference." Just as the president of the Royal Geographical Society suggested that the existence of a neutral zone or a frontier was nothing
more than a "standing invitation to a strong and highly civilized power"

to push forward until this neutral zone had been absorbed, so *Kim's* pretense of a neutral zone of friendship between Kim and the lama is, in reality, nothing more than a means to absorb an object of knowledge and thus own it.

This ideology of absorption is expressed at the beginning of the text, when the lama travels to the Wonder House (Lahore Museum) to seek directions to the River of the Arrow. This museum is run by an Anglo-Indian, whom the lama addresses as a "Fountain of Wisdom" because of the documents he owns about Buddhism: "For the first time he [the lama] heard of the labours of European scholars, who by the help of these and a hundred other documents have identified the Holy Places of Buddhism." Interestingly, the lama had apparently marched for four months in order to see a picture of his own monastery, for the curator is unable to reveal the location of the River of the Arrow. What the lama gains, instead, is a reverence for the methods of representation of European scholars; having witnessed a "mighty map" and a photograph, he has gained a lesson in the superiority of European methods of capturing and storing information (26). This so impresses the lama that he gives money to send Kim to school, claiming that he wants Kim to be like the man at the Wonder House. "That is my hope," he says, "for he was a Fountain of Wisdom—wiser than many abbots" (130). Indeed, Kipling appears to stress the European superiority in knowing the Orient—by virtue of superior vision not clouded by superstition—in the metaphor of the curator's glasses. When the lama appears at the Wonder House, his glasses are scratched to the point of obscuring his vision; the curator immediately notices this and gives the lama his own glasses, which he claims are made of "crystal and will never scratch." The metaphorical value of this gesture is expressed in the next statement, "May they help thee to thy River"—as if enlightenment can only come through embracing the vision of the Europeans (29).

It is in this sense that the "reality effect" takes place within the novel. Europeans, who are fighting off the superstition and clouded vision of India, bring reality to India. In this sense, *Kim* is a realistic novel, for it is about overcoming the fabulous effects of India. The nothingness of Kim's identity is not coincidentally related to his photographic vision; only through the elimination of subjective reality can the truth reveal itself. Kipling not only wants to create a realistic picture of India; he wants to define realism. By turning Kim into a camera of sorts, he plays upon the European prejudice toward a belief in objective knowledge—in this sense the entire novel turns into a project of transforming clouded Indian knowledge into objective European knowledge, with pictures of itself.

When Colonel Creighton begins to train Kim to be a surveyor for the military, he explains the goal of Kim's training: "Thou must learn how to make pictures of roads and mountains and rivers—to carry these pictures in thy eye till a suitable time comes to set them upon paper. Perhaps some day, when thou art a chain-man, I may say to thee when we are working together: 'Go across those hills and see what lies beyond'" (126). Trained to be a chain-man, Kim must ironically travel without a chain because of hostility toward surveyors in the frontier areas surrounding India. Clements Markham expressed the real danger of carrying chains into the Himalayas for surveying purposes: "The chain is not generally used by the Topographical Surveyor, because it is politically obnoxious to independent tribes, and is looked upon as the sure harbinger of loss of territory."[25] The issue of surveying, then, should be seen not as an objective pursuit of knowledge, but as a political contest for territory. Since chains could not be used in independent territories, rosary beads and pacing were used to measure distances instead, by surveyors in disguise. During a visit to Mahbub Ali, Kim is "ordered" to "make a map of that wild, walled city." And, "since Mohammedan horse-boys and pipe-tenders are not expected to drag Survey-chains round the capital of an independent native State, Kim was forced to pace all his distances by means of a bead rosary" (175). With the "help of his little Survey paint-box of six colour-cakes and brushes," Kim is also able to sketch a map of the city for Creighton. Mahbub Ali explains that the map "must hold everything that thou hast seen or touched or considered"—in this sense, it is supposed to come as close to photographic realism as possible (174).

Kim uses European rationality to overcome the fabulist Indian vision when Lurgan, a shopkeeper who "does magic" and boards Kim during his school break, attempts to hypnotize him. At first Kim succumbs to Lurgan's hypnotic powers but then learns to overcome them. Kipling describes this process: "So far Kim had been thinking in Hindi, but a tremor came on him, and with an effort like that of a swimmer before sharks . . . his mind leaped up from a darkness that was swallowing it and took refuge in—the multiplication-table in English!" (159). Multiplication, which can rescue you from hypnotic powers, is also described as the way to accurately map a territory without a measuring chain. Hurree explains that Kim will be able to use multiplication by carrying a rosary, as he claims:

> A boy who had passed his examination . . . could, by merely marching over a country with a compass and a level and a straight eye, carry away a picture of that country which might be sold for large sums in coined silver. But as it was occasionally inexpedient to carry about measuring-chains, a boy would do well to know the precise length of his own foot-pace. . . . To keep count of thousands of paces, Hurree Chunder's experience had shown him nothing more valuable than a rosary of eighty-one or a hundred and eight beads, for "it was divisible and sub-divisible into many multiples and sub-multiples." (169)

By learning the powers of memorization and math, Kim is able to create a picture of India that is more real than any native could create. However, it is Kim's own Irish identity that may give him the power to see and memorize India. Standing between England and India, Kim's ambivalent or marginal identity in the British Empire enables him to befriend the Indians who can provide him with information.

The revelation of Kim's racial identity is interestingly placed, without explanation, next to the sentence about possessing the lama: "The lama was his trove, and he purposed to take possession. Kim's mother had been Irish, too." While we had earlier learned that Kim's father had been Irish, it is only at this moment that we learn that Kim is Irish through and through. Strangely, Kim is called English in the first paragraph, when he "sat, in defiance of municipal orders, astride the gun Zam-Zammah." Kipling explains: "There was some justification for Kim . . . since the English held the Punjab and Kim was English. Though he was burned black as any native . . . Kim was white" (19). Though Kim's identity is somewhat confused between English and Irish, black and white, it may be that his very Irish background allows him to eliminate his identity and absorb India. In 1866, Montgomerie described the unique role the

Scottish played in surveys of Afghanistan. A Scottish colonel, Johnstone, told Montgomerie a story of a time when a member of the Afridi tribe saw his Scottish crest, the "flying spur." Montgomerie writes:

> The Afridi asked its meaning, and was told that, in former days, men in Scotland were as lawless as the Afridis, and that when the larder was empty, a dish was put before the chief, containing only a spur with a pair of feathers fastened to it being a signal to him and his followers to boot and spur, and be off to raid the cattle over the border, and that the "flying spur" then became the badge worn by the clan. The hearts of the frontier Afridis warmed to the Colonel, when they found he was the descendent of the British Afridi.[26]

As an Irishman, Kim's perceived role as a British Afridi may have contributed to Kipling's construction of him as a "friend of all the world." The fact that he is "burned black as any native" may metaphorically stand for the fact that the British often described the Irish as black in the nineteenth century. More specifically, the relation between being Irish and claiming the lama as his "trove" could suggest an inherent greediness in the Irish—it could also, however, suggest that the Irish, with a less stable identity than the English, were more capable of emulating that nothingness required of the photographic eye. Marginalized from the center of empire, the Irish become marginalized from themselves. The kind of uncertainty that Kim continually expresses over his identity is never expressed by the English characters of the novel. The Irish, who are closer to the Tibetans or Afghans, are better suited to take possession of these countries for England.

THE FORWARD SCHOOL

Britain's need for objective realism is undermined by the terror of the very people that they see as being in a better position to perceive it. While giving glasses to a lama may appear to "clarify" his view, the lama is needed only to aid in destroying other lamas in Tibet. His position, then, is always an ambivalent one—his eyes are needed only to see the invading Russians or Chinese. Britain's fear of Russian aggression is evident at the beginning of *Kim,* when we learn that "Five Kings" in the North had confederated, "who had no business to confederate"—and the fear is that they are in league with Russia against Britain. This leads the British—under the command of Creighton—to organize a war against these kings and subdue them. But where *Kim* begins to read as propaganda for a forward frontier policy is in the next episode of this military saga. Hurree Babu explains: "But the war was not pushed. That is the Government custom. The troops were recalled because the Government believed the Five Kings were cowed; and it is not cheap to feed men among the high passes." (38). Instead, two men were left to guard the passes and monitor the building of roads in the Himalayas. Hurree, a Bengali, was hired to pay the "coolies" building the roads, and he claimed: "I sent word many times that these two Kings were sold to the North; and Mahbub Ali, who was yet farther north, amply confirmed it. . . . I sent word that the roads for which I was paying money to the diggers were being made for the feet of strangers and enemies. . . . For the Russians" (222). Kim's ultimate success as an agent of the Secret Service is proven by his finding the "love-letters to the Czar" from these two northern kings, Hilás and Bunár (223). The results of his effort change the course of British history in India, as these independent kings are instantly deposed by the British: "The British Government will change the succession in Hilás and Bunár, and nominate new heirs to the throne" (275). Kim's success, then, comes about in correcting the errors of a weak frontier policy.

Kipling artificially constructs hybridism as a necessary byproduct of espionage, which—in the end—can never be anything more than a military strategy. Kim is described in the novel as "a Sahib and the son of a Sahib. . . . But no white man knows the land and the customs of the land as thou knowest" (101). Similarly, Hurree Babu is described by the Russian: "He represents *in petto* India in transition—the monstrous hybridism of East and West" (238). There are several levels of hybridity within the text: Kim, as Irish-English, serves as a revision of Singh's journeys into the Himalayas, turning Singh into a white man, though Kim is "black as any native." In this gesture alone, the ambivalence regarding native discovery is revealed. The pundit/native becomes pundit/white, rescuing the work of Singh for Montgomerie. On the other hand, Hurree's credibility becomes questionable by the end of the narrative. As Kim becomes more white within the narrative, Hurree becomes dismissed as duplicitous. After Hurree helps Kim to deceive the Russians and steal their maps, Kim turns against him: " 'He robbed them,' thought

Kim, forgetting his own share in the game. 'He tricked them. He lied to them like a Bengali. They give him a *chit* [a testimonial]. He makes them a mock at the risk of his own life' " (277). While Hurree was working for Kim, he also set about getting a letter of recommendation from the Russian, before he was killed. And it this one deceptive maneuver— though the entire narrative valorizes deception—that turns Kim against Hurree.

Interestingly, it is Hurree who continues to believe that the great game must continue, long after others think the problem of allegiances with Russia has been resolved. He claims: "When every one is dead the Great Game is finished. Not before" (222). On the one hand, this statement may be read as an overzealous commitment to the Secret Service of England; on the other hand, it may be seen as a very cynical expression of the fact that England's "game" is killing everyone. And yet, there is a third possible reading of this statement, which may be more clearly understood by returning to the words of the president of the Royal Geographical Society in describing Afghanistan:

> The distinctive characteristics of the tribes inhabiting this wild borderland are selfishness, vanity, treachery, vindictiveness, and general lawlessness. Their attitude towards one another is one of thinly-veiled antagonism which may at any moment break out into open hostility. This inter-tribal aggressiveness is only overruled by . . . the fear and hatred excited by the smallest suspicion of foreign interference, so that it may be said of the Afghans of the frontier that they are never at peace except when they are at war. Asked by a British officer what would be their attitude in the event of war between Great Britain and Russia, a party of tribesman answered: "We would just sit up here on our mountain tops watching you both fight, until we saw one or the other of you utterly defeated; then we would come down and loot the vanquished to the last mule! God is great! What a time that would be for us!"[27]

The great game—in this sense—will not be over until the Russians and British are dead, and Hurree can collect the loot. And while Kim's acts of deception are always for England, Hurree remains in the ambivalent position of a man with potentially no loyalties. If England goes forward and dies, Kipling suggests, it will be because of men like Hurree and

Mahbub Ali. And yet, it is precisely because of the fear of these men that England must go forward.

RACIAL FIXITY

"When every one is dead" the game is over, and the narrative itself mimics this goal of elimination. If the pundit must be used, he will not interfere; in fact, he can be translated later into a British subject. But the true goal of elimination is what is perceived to be the native voice itself—the fabular Oriental. This perspective will be overwritten with the voice of the enlightened Englishman. If every point on the earth is said to have its own special position in terms of latitude and longitude, the true significance of this perspective is the belief that "no other point can interfere."[28] The earth can be overwritten with the one true law of geometric order. The problem, of course, with this perspective is that there are still Indians pulling up survey stakes and disloyal Afghans and Tibetans who close the borders of trade. So for the great game truly to be finished, all human interference with the points of latitude and longitude must be destroyed—"every one" must be dead.

While JanMohamed may claim that Kipling celebrates difference, Mark Paffard writes, "Time and again Kipling draws Indian characters so deftly that they remain in the memory, but does so explicitly as *types.*"[29] Patrick Williams also locates the stereotypes in *Kim,* suggesting that the stereotype is "perhaps the principal mechanism in ideologies of discrimination and domination at work in colonialism."[30] However, by examining the supposed fixity of the stereotype, we may perhaps begin to understand the relation between cartographic fixity and racial flexibility. Homi Bhabha claims that the stereotype is not "the setting up of a false image which becomes the scapegoat of discriminatory practices." Instead, he proposes, "it is a much more ambivalent text of projection and introjection, metaphoric and metonymic strategies, displacement, overdetermination, guilt, aggressivity. . . . It is the scenario of colonial fantasy which, in staging the ambivalence of desire, articulates the demand for the Negro which the Negro disrupts."[31] *Kim* is a text that enables the examination of the productivity and ambivalence of stereotypes. The stereotypes in *Kim* can be gauged according to their mixture with being English—Irish-English, Afghan-English, Bengali-English. *Kim* explores the possibility of the Other working for the English; but as the narrative progresses, it is only the white man who may still have the potential to fully evolve into an Englishman. The dangerous dou-blespeak between racial categories, however, is only made possible by the absolute fixity of geographical coordinates in *Kim*. It may be, in fact, that the absolute location of spatial coordinates in India con-tributed to more flexibility in the portrayals of individuals, who are then safely confined by these boundaries. Not only are they contained, how-ever, they are also made readable in London; territory is certain, fixed, and communicable.

As geographers carried the chain, or rosary, into Afghanistan and Tibet, the absolute and objective nature of their work—creating coordi-nates for a map—allowed them to overlook cultural or ethnic bound-aries. The permanent elimination of those very boundaries for the sake of trade would in turn become the ultimate goal. In 1904, the geogra-pher Francis Edward Younghusband and his army massacred close to six hundred Tibetans, forcing the Dalai Lama into exile, in an attempt to secure a trading route to China for the British. For his efforts, Younghus-band was knighted and a trading agreement was signed between Tibet and Britain in Lhasa. The Chinese, refusing to honor this agreement, obliged Britain to sign a treaty declaring Chinese rule in Tibet.[32] Encouraged by the latter agreement, China would violently invade Tibet in 1906 for the first time in ten centuries, shooting into unarmed crowds in Lhasa. After continual sovereignty disputes between China and Tibet, as many as 430,000 Tibetans were killed by the Chinese between 1949 and 1959. In 1994, however, the British still stood by the decision they had made at the beginning of the century. The British government announced its official position on Tibet: "Independence for Tibet is not a realistic option. Tibet has never been internationally recognised as an independent state, and no state regards Tibet as independent."[33] This inability to deal with the possibility of independence, indeed, is written into the history of both Afghanistan and Tibet—and has shaped the his-tory of mapping itself, which produced narratives of racial ambivalence only to later neutralize them through their separation from the final product of the map. Montgomerie wrote of evidence on the map that geographers had "been 'nibbling' at Tibet in all directions" long before the pundit explorations.[34] Regardless of the slipperiness of racial rela-tions, the certainty of the scientific project of mapping would lead the

English to see their role in the Himalayas as incontrovertible. It would also lead to the renaming of the pundits with letters or numbers, in this way refixing their identity with the absolute coordinates of the map—an area that could never be questioned. But underneath those coordinates, the game continued, though it may never have been quite the game imagined by Europe. Rather than a dialogue between Russia and Britain, the game would be perpetually interrupted by the ghostly presence of a third space that produced racial ambiguity. The unimaginable nature of this game would serve to haunt the borders to which the British were perpetually drawn.

PART TWO

Aerial Photography

CHAPTER 3

The Gendered Aerial Perspective

IN 1858, Gaspard Félix Tournachon, also known as Nadar, took the first aerial photograph from a tethered balloon over Petit Bicêtre. In the same year, Aimé Laussedat—who had earlier attempted to launch a camera from a kite—began photographing Paris from rooftops and steeples around the city. While Tournachon is credited with taking the first aerial photograph, Laussedat is credited with developing the field of photogrammetry, or the use of photographs in cartography.[1] However, because balloons and kites were too unstable for further development in this field, it wasn't until the development of the airplane in 1903 at Kitty Hawk that aerial photogrammetry became plausible. By 1915, the French were mass-producing the first cameras designed specifically for aerial use, to be used for aerial reconnaissance work during World War I.[2] During the 1920s and 1930s, cameras were mounted on airplanes to conduct surveys that took overlapping vertical photographs. This new form of surveying was called photogrammetry and was eventually to replace traditional triangulation surveys.[3]

In May of 1919, aerial photography became commercialized in London, when Aerofilms was registered by F. L. Wills and Claud Graham-White—recently returned from the war—as the first aerial photography company.[4] The airplane, which Wills and Graham-White began to use to map London in 1921, had become more than a spectacle of military heroism; it had become an urban reality, and aerial photographs could represent the public to themselves in a disorientingly original way. When the aerial photograph became popular, promoters were optimistic about its effect upon the public: "As the Public continue to see pictures of the earth from above, and aerial maps or pictures are continually being used, the world becomes more air-minded." This in turn, it was explained, would help the development of aviation.[5] This belief in air-mindedness was called the "winged gospel" of the time, "a secular creed about the promise of the future which was promulgated with the intensity of an evangelical religion," as Susan Ware explains. "In this worldview, aviation would reorder human society and promote world peace by breaking down isolation and distrust. When everyone took to the air, society would be transformed along the lines of democracy, freedom, and equality—and perhaps even women's liberation."[6]

In 1925, Virginia Woolf called the airplane a symbol "of man's soul." In *Mrs. Dalloway,* she wrote: "Away and away the aeroplane shot, till it was nothing but a bright spark; an aspiration; a concentration; a symbol . . . of man's soul; of his determination . . . to get outside his body, beyond his house, by means of thought, Einstein, speculation, mathematics, the Mendelian theory—away the aeroplane shot." Getting outside the body with the aid of an airplane was no longer a perspective reserved for pilots; it was, however, still reserved for men, a problem that is played out throughout *Mrs. Dalloway.* At the beginning of the novel, Clarissa Dalloway imagines "the high singing of some aeroplane overhead was what she loved; life; London; this moment of June." Virginia Woolf suggests that there is an "unseen part of us, which spreads wide"—a part that Clarissa attempts to bring together by throwing a party.[7] Just as Clarissa tries to draw herself together into one center, so an airplane draws the city together.

Set in the streets of London, just after World War I, the novel introduces characters in the moment that they look up at an airplane; indeed, as a kind of roaming omniscient narrator, the airplane appears to determine the logic of the narrative itself. As people gather at Buckingham Palace, hoping to see the queen, "the sound of an aeroplane bored ominously into the ears of the crowd. . . . Every one looked up." The airplane is a means of getting perspective, of getting beyond one's house and body and escaping into pure thought. The airplane draws out the longings of the characters for distant places ("Ah, but that aeroplane! Hadn't Mrs. Dempster always longed to see foreign parts?"), while also becoming a kind of fulcrum that draws together the disparate elements of urban existence.[8]

Similarly, in Woolf's "Kew Gardens," the position of the narrator mimics an aerial perspective—sudden shifts in altitude and visual resolution are the dominant traits of this narrative style. From closely viewing the movement of insects to suddenly seeing people as blurs on the horizon, the shifting altitude could be said to mimic that of an airplane dipping and circling over Kew Gardens. And at the end of the story, the narrator concludes, "In the drone of the aeroplane the voice of the summer sky murmured its fierce soul." The voice that ultimately tells the story of a summer's day in the park is that of the airplane, which makes out forms and colors more than details: "Yellow and black, pink and snow white, shapes of all these colours, men, women, and children were spotted for a second upon the horizon, and then . . . they wavered and sought shade beneath the trees, dissolving like drops of water in the yellow and green atmosphere, staining it faintly with red and blue."[9] In *Mrs. Dalloway,* the narrative perspective similarly dives upon the intimate conversations of couples in a park before moving on to the next object of observation. The question, then, becomes one of interpretation: how can one read the seemingly disconnected but poignant stories happening all over London? And who—man or woman—has the power to make this interpretive leap?

Shortly after the start of the World War I, the Germans began to photograph the entire western front every two weeks, and by the end of the war, they were taking approximately four thousand aerial photographs a day. The volume of information was immense, but the ability to read these photos was often lacking. For instance, differences in shadows on an aerial photograph could make it next to impossible to recognize the same site in two photographs taken at different times. An aerial photography school was established in the United States shortly after the war ended to interpret aerial photographs; the school was quickly in the position of having to deal with large shipments of photographs every two weeks from the British, Americans, and French.[10] The enemy's moves had to be interpreted as well as recorded: "The interpretation of aerial photographs was, of course, a real problem and one of vast importance. . . . The smallest detail had to be accounted for in terms of military importance."[11] The notion that geographic information must be understood, stored, and preserved led to the institutionalization and perpetuation of warfare photographic interpretation long after the end of the war.

Warfare had formulated, in many ways, the very meaning of the aerial perspective. Lowell Thomas wrote in 1927, "The public mind has come to associate the airplane with death and destruction."[12] Paul Virilio, in *War and Cinema,* explains this shift: "On board an aeroplane,

the camera's peep-hole served as an indirect sighting device comple-
menting those attached to the weapons of mass destruction. It thus pre-
figured a symptomatic shift in target-location and a growing
derealization of military engagement." It is precisely because what is sited
is a target that these sites are viewed as objects of loss. "*What is perceived
is already lost,*"Virilio writes.[13] Warfare had led to the establishment of an
industry that saw its subjects as hostile, dehumanized, and disorganized.
Indeed, territory could only be correctly and safely interpreted during
war through the view from above—the enemy was avoided and his
threat was neutralized through the lens of a camera. Postwar aviation
would ironically have to overcome the public's perception that the airplane signified death and destruction, while at the same time building up
the possibility of future wars to justify its continued development.

Lowell Thomas, the journalist and radio announcer chosen to
accompany the first flight around the world in 1924, explained this connection: "All commercial planes can be converted into engines of war
at a moment's notice; big passenger and freight planes can be immediately used for bombing and transporting troops." Thomas, in 1926, took
his family on a tour of the "European skyways," partly in an attempt to
prove that airplanes were safe for tourism. Ironically, Thomas himself
often does not appear to be able to see scenes other than war from his
aerial tour of Europe. As he passes over the site of the third battle of
Ypres, he thinks: "As we look down the hilltop is green and quiet, but
that day the horrors of hell were loosed on it. A rainstorm broke and
the British advanced in a sea of mud. And now the German cannon
blasted them, and deluged them with gas. They struggled up that now
pleasant slope, through a poison-filled bog."[14] The confusion of tenses
in the ambiguous use of "now" in this passage demonstrates a kind of
slippage between war—traumatically imprinted in the memory—and
peace.

Thomas, in *European Skyways,* captures that perpetual displacement
between military and civilian technologies. The civilian airplane
becomes seen as a temporarily camouflaged military bomber, just as a
"pleasant slope" is also a "poison-filled bog." Aerial photography allowed
the postwar civilian to see evidence of both, as the pockmarked landscape carried perpetual memories of war. Thomas describes this sensation of perceptual doubling: "As I speed across the sky the whole history

of the region below me, or such fragments as I can recall, come rushing to mind. So I really see the double panorama—one with my eyes, the other with my imagination." Interestingly, while Thomas unceasingly recognizes battle sites and reports military history on this family holiday, his wife is—by turns—frightened, airsick, and bored. She alternatively practices inflating her life jacket, reading, and feeling sick—and certainly does not see what her husband has seen; he complains of her attitude: "so womanlike (?) she had to do her worrying out loud."[15] The militaristic aerial perspective, then, appears to be a distinctly gendered one; yet the curiously placed question mark after "womanlike" suggests an ambiguity, or even confusion, surrounding these very gender roles.

THE ANDROGYNOUS PILOT

The gendering of aerial photography as a military masculine enterprise relates directly to the history of women in aviation. In 1910, a French woman, Elise Baronne de Laroche, became the first licensed woman pilot. In 1911, Harriet Quimby became the first licensed woman pilot in America, and in 1912, the first woman to fly—in a "specially designed, hooded purple satin flying suit"—across the English Channel. Women's early role in aviation has been described as emerging "parallel to the Ziegfield Follies, common burlesque, or the circus, all popular—and extremely sexist—forms of entertainment at the time." Wing-walkers, parachute jumpers, and stunt flyers were the common roles for women in aviation; and because of their role as entertainers, it was also considered essential that they look the part: "diminutive stature, general adolescent appearance, very feminine clothing, and the ability to look helpless."[16] Aviation was an entertainment business that capitalized upon placing women in Pauline-like danger, setting a trend for fashion consciousness in female pilots that would continue for decades.

Before the advent of airplanes, women were commonly sent up in balloons. Amelia Earhart writes, with a certain nostalgia: "The beplumed and beribboned equipages were designed to harmonize with and enhance the appearances of the performers. And vice versa. . . . It seems almost as if the spectacular side of aerial entertainment has never reached so high a pinnacle as it did during this fabulous period of balloon pageantry" (fig. 10).[17] Margie Hobbs, alias "Ethel Dare, the Flying Witch," was famous for her use of rope ladders in 1919 and 1920. Dean

M. S. BLANCHARD celebre aeronauta
al momento del volo aerostatico da lei eseguito in Milano
in presenza delle LL. AA. I. I. e R. R.
la sera del 15 Agosto 1811.

10. Madame Blanchard, the first woman aeronaut, was also the first woman to lose her life while flying. In 1819, she fell out of her basket when her balloon caught on fire. Photograph from the Aviation History Special Collections, Harvey Mudd College.

Jaros writes, "She was a former circus trapeze artist who well understood the draw of scanty attire and well-posed photographs."[18]

If women were allowed into the field of aviation when it was understood as a form of entertainment, their role in aviation would be continually suppressed during periods of war. Pilots like Alys McKey, Ruth Law, and Katherine Stinson were turned down, despite repeated applications, for combat pilot positions during World War I. Women, while allowed to play the role of aerial spectacle, had more difficulty becoming accepted simply as pilots. Amelia Earhart, describing her frustration at women aviators being called "Ladybirds," "Sweethearts of the Air," or "Powder Puff Pilots," writes, "We are still trying to get ourselves called just 'pilots.'"[19] Because women were pushed out of the aviation industry during wartime, they also were excluded from the field of aerial photography, which came to dominate cartography during and after World War II.

Interestingly, women often adopted short hair and military attire in order to look inconspicuous in aviation circles. Amelia Earhart describes "snipping inches off my hair secretly" when she became a pilot. "In 1920," she writes, "it was very odd for a woman to fly, and I had tried to remain as normal as possible in looks, in order to offset the usual criticism of my behavior." In this passage, the "normal" is ambiguous; is it "normal" to have short hair or long hair as a female pilot? Women were caught in that ambiguous space between wanting to maintain their femininity so as not to appear "crazy" and wanting to cut their hair or otherwise dress in masculine attire so as not to stand out as women. "Flyers dressed the part in semi-military outfits," Earhart writes, "and in order to be as inconspicuous as possible, I fell into the same style."[20] World War I would lead to a rise in this semi-military, masculine style among women pilots, in contrast to their beplumed and beribboned history.

Bessie Coleman, in 1921 the first African American woman to receive a pilot's license, further complicated this issue of style. Coleman, who was close to half Choctaw Indian, faced racism as well as sexism in her pursuit of flying. Because U.S. flight schools would not allow blacks (or often women) to attend, she moved to France and received an international pilot's license from the Fédération Aéronautique Internationale. Back in the United States, however, Coleman was unable to find a job or to purchase a plane because of discrimination. She often had to borrow or rent planes in order to perform her barnstorming acts across the country. Because of her sense of outsideness as a black women, Coleman began to publicly associate herself with Joan of Arc, claiming that she studied "in the city where Joan of Arc was held prisoner by the English."[21] Joan of Arc, who adopted male battle dress, fought to free the French from the English. Coleman saw herself in a similarly androgynous role in which she was fighting for African American rights against Anglo-American tyranny. Her goal, she claimed, was "to make Uncle Tom's Cabin into a hangar."[22] "So determined she was," a friend said, "to make Negro men become air-minded."[23] The use of the gendered "men" in this statement may reveal that Coleman was more interested in overcoming racism than sexism. Her continued fascination with Joan of Arc, however, demonstrated her awareness of her complex role as a woman. Cementing her affinity with Joan of Arc, Coleman chose to wear French military uniforms for her performances (fig. 11).

World War I, by excluding women from the air, may have pushed women into adopting masculine attire in order to hide their sex. Before

11. Bessie Coleman, in her military-themed uniform, before she fell to her death from her plane. Photograph from the New York Public Library.

World War I, femininity was generally accentuated in order to draw attention to women as performers; after the war, it appeared more often than not to be diminished. Simultaneously, aviation became more identified as a male sport after the celebrity of the wartime flying aces. Women who dreamed of being pilots would often dream of being men. Virginia Woolf, in *The Voyage Out,* describes the complex relation between women and planes in a scene where Mrs. Thornby over tea reveals her desire to fly: "It was odd to look at the little elderly lady, in her grey coat and skirt, with a sandwich in her hand, her eyes lighting up with zeal as she imagined herself a young man in an aeroplane. For some reason, however, the talk did not run easily after this."[24] In *Mrs. Dalloway,* Mrs. Dempster also may "long for foreign parts" when she sees a plane; but she recognizes, at the same time, that the very "Mrs." attached to her name will forever keep her from flying. Instead, seeing the plane, she wistfully imagines that "there's a fine young feller aboard it."[25] Virginia Woolf repeatedly describes women as men when they think of flying.

By the late 1920s, the role of women in aviation had again shifted, slightly, and women were in demand as passengers and pilots in order to demonstrate that flying was safe enough for women. Suddenly, the era of the barnstormer and the flying ace was viewed as detrimental to the promotion of aviation as a commercial enterprise. "Cult figure status for the aviator had become counterproductive for business enterprise," Dean Jaros explained. "American enterprise sought to tame the image of aviation by incorporating women into it. The underlying assumption was that women are inherently timid, unwilling to take risks, indecisive, uncoordinated, physically weak, and scatterbrained. If *they* can pilot airplanes . . . aviation must be safe and piloting must be easy." Amelia Earhart, in 1928, entered aviation in such a capacity, as a passenger on a transatlantic flight that was geared to naturalize the image of women's long-distance flying. Earhart commented, "From the beginning it was evident that the accident of sex—the fact that I happened to be the first woman to have made the Atlantic flight—made me the chief performer in our particular sideshow."[26] Complicating Earhart's "accident of sex" was the rumor that she was chosen for this flight because she looked like a man: Charles Lindbergh. Boston papers immediately called her a "veritable girl Lindbergh," the French air minister called her "Lindbergh's double," and portrait artist Brynjulf Strandenaes claimed, "She looks more like Lindbergh than Lindbergh himself."[27] Because of what many perceived as an uncanny resemblance, Earhart would be instantly dubbed "Lady Lindy."

After the flight, Earhart captioned a photo of herself with the two pilots: "Two musketeers and—what is a feminine musketeer?" Earhart's own gender confusion in this caption would become a predominant theme throughout her career, leading her to be called "America's first androgynous sex symbol."[28] Her choice to adopt masculine attire, while a common fashion trend among female aviators, complicated her "job," which she described as "to sell flying to women."[29] Interestingly, she would later adopt women's clothes as a disguise when she wanted to travel incognito. Amelia Earhart *disguised* as a woman is perhaps a fitting symbol for the complex role that women in aviation found themselves in between wars.

Women were commonly drawn to aviation for the feeling of freedom that it provided from the limited roles available to them as women. In 1930, aviator Margery Brown exclaimed: "Women are seeking freedom. Freedom in the skies! . . . The woman at the wash-tub, the sewing machine, the office-desk, and the typewriter can glance up from the window when she hears the rhythmic hum of a motor overhead, and say, 'If it's a woman she is helping free me, too!' "[30] Some women, like Edna Gardner Whyte, associated flying with escaping the limited roles available to women, as well as their oppression: "I had lived for eighteen years before women were 'given' the right to vote. . . . I specifically chose to enter aviation, a man's field in a man's world spurred by a fierce desire to *be* someone." Margery Brown and Edna Gardner Whyte associated flying with escaping their own confined gender roles (the wash-tub, the sewing machine, the typewriter) and entering those assigned to men. Often, "being someone" meant leaving behind men's chivalrous attention to embrace independence. Brown wrote: " 'Let me alone! I don't need any help. I can look after myself.' This is just what a woman does say . . . when she gets into a plane, opens the throttle, tosses her head, and goes off sailing alone into the sky."[31] Flying promised to provide a practical end to the economic dependence of women, as well as entry to an industry that had been dominated by men during the war. Unfortunately, the entrance of women into the world of aviation was predicated upon women selling themselves as the weaker sex. The idea that "if women can fly, anyone can" meant that women were forced to contribute to the stereotype of their own helplessness in order to enter aviation.

Earhart's job—convincing women that they had the freedom and ability to fly—therefore, was complicated by the simultaneous promotion of the idea that women were less able, more at risk, and less qualified to take to the air. Anne Morrow Lindbergh, after reading about the qualities required to be a pilot, commented that "they are comically so opposite to anything I have: *Instantaneous co-ordination between his muscles and thoughts . . . calm and levelheaded at all times.*" She continued, "To all of which I sigh, '*Rien à faire.*' Good eyesight is my only qualification."[32] Women like Amelia Earhart and, before her, Harriet Quimby, struggled to help women overcome feelings of inferiority in order to become pilots, only to find that women were constantly restricted from entering the profession. Quimby wrote in 1912: "Flying is a fine, dignified sport for women, healthful and stimulating to the mind. . . . I see no reason why they cannot realize handsome incomes by carrying passengers . . . from taking photographs from above, or from conducting schools of flying."[33] The very thing that Quimby and Earhart tried to convince women was a means to their economic and emotional freedom would be repeatedly made an impossibility by a distinctly male-dominated industry.

Though women were allowed to enter aviation when the gender stereotypes surrounding them proved useful to the industry, they were repeatedly dropped when they appeared to be interfering with men's control over it. World War I and World War II drastically curtailed women's participation in the field of aviation. After World War I, the stunt pilot Ruth Law was "informed" by her husband that she could no longer fly. She explained the consequences: "I kept my nerves under control for two years and then I had a nervous breakdown. . . . Sometimes the sight and sound of planes overhead is maddening, but my flying days are over" (121). Though a small group of women, known as WASPs (Women Air-force Service Pilots), was allowed to work in transport flying during World War II, these same women were barred from the air force after the war. General Henry H. Arnold adamantly denied their requests: "You will notify all concerned that there will be no—repeat—no women pilots in any capacity in the Air Force after December 20" (125). Many of these

WASPs, blackballed by the aviation industry as well, fell into despondency or even killed themselves (115–148). Claiming to feel "disappointed in everything," Mary Wiggins committed suicide shortly after the war ended. Helen Richey committed suicide two years after the war ended, unable to find employment either in commercial or military aviation. Ruth Nichols died in 1960, her death also ruled a suicide. Others simply disappeared into domestic life. Nancy Harkness Love settled into a kind of depression, and "Love's behavioral pattern was repeated among hundreds of other WASPs" (127–130).

In Canada and Britain, female pilots faced, possibly, an even greater sense of rejection during and after World War II. The motto of the Women's Division of the Royal Canadian Air Force during World War II was "We serve that men may fly." Florence Elliott Ward, who received a pilot's license and applied to the RCAF, explained: "They gave me a flat 'no' unless I was willing to sit and type, thus releasing some man to fly. I thought, 'In a pig's eye'" (40). Daphne Paterson, who trained to become a flight instructor, reported her experience: "I was told there was no place for female instructors with the British Commonwealth Air Training Plan." After the war ended, the situation did not improve. Joyce Bond wrote, "I knew at war's end, with all those returning airmen, I stood very little chance of being hired as a pilot and I was not prepared to remain on the edge of flying any longer." Jessica Jarvis, another pilot, believed that "women were not wanted on the field by some of the men because they detracted from the mystique of the male pilot." Marian Orr expressed her despair at this realization: "I felt so empty. It was as if my whole life was behind me. . . . I had all that experience and I knew I couldn't put it to use."[34] The promotion of the stewardess as a kind of flying nurse/sex symbol (they had to be registered nurses and could not be married) after the war would seal women's exclusion from the role of pilot for the next thirty years.

It was in this way that the promise that women like Amelia Earhart offered to women for a decade began to vanish by the late 1930s and disappeared almost completely after the beginning of World War II. "'Domesticating' aviation helped to sell it," Dean Jaros wrote. But by the late 1930s, aviation had been generally accepted as a part of modern life. Jaros continued: "The point had been made. . . . The airwomen had worked themselves out of a job."[35] Their job, which was simply being women who could fly, no longer seemed to mesh with the goals of a male-dominated aviation industry.

Domestic Aerial Photography

Women's role in domesticating aviation in the late 1920s and early 1930s correlated with the part they played in promoting aerial photography. During the 1930s, husband-and-wife flying teams, such as the Lindberghs, Haizlips, and Thomases, were especially popular, embodying the ease of making a "home in the air." Women were often the family photographers in these situations, filling the role of the family vacation traveler taking photos. On the one hand, the photos themselves were useful in promoting aviation; on the other, the image of women with cameras may have been even more successful in domesticating aviation. Amelia Earhart, applauding Anne Lindbergh's teamwork with husband Charles, commented: "He increasingly counts on her active cooperation. One of her responsibilities is photographing; or she may take the controls when her husband is preoccupied with 'shooting the sun.' "[36] Anne Lindbergh snapped photos on extended trips that she and her husband took to popularize aviation and promote air travel; two of her books, *North to the Orient* (1935) and *Listen! The Wind* (1938), included photographs and functioned as adventure travel narratives in seemingly exotic places. Because of the Lindberghs' popularity as a married flying couple, Anne Lindbergh's books also serve to promote the flying marriage itself as a kind of endless honeymoon. Richard and Mary Light similarly engaged in aerial photography in Africa in the 1930s, discovering Mount Stanley from the air in 1937. Their photographic work, combined with endless tales of exotica, made flying look at once romantic and adventurous, and yet no more dangerous than a family picnic.

In the early 1920s, women were sometimes pilots for photographers needing aerial shots. Bessie Coleman, for instance, flew photographers from *Pathé News* in both Germany and America.[37] In circumstances such as these, there appeared to be no overtly gendered associations between flying and photography. In 1932, I. N. Phelps Stokes and Daniel C. Haskell said that the demand of the public for urban aerial views seemed "to have become a universal hobby—almost a mania."[38] The mania for the aerial perspective also influenced female photographers of the 1920s and 1930s. Photographer Margaret Bourke-White, for instance, exclaimed,

12. Margaret Bourke-White (here, outside her studio at the Chrysler Building) was famous for her daring vantage points. Photograph by Oscar Graubner/TimePix.

"Airplanes to me were always a religion."[39] Bourke-White's studio was on the sixty-first floor of the Chrysler Building in Manhattan, where she often experimented with the aerial perspective (fig. 12).

It wasn't until the idea of domesticating aviation came into being, however, that photography began to be seen as the realm of women passengers. Amelia Earhart was the photographer on her first passenger flight across the Atlantic, taking photos while lying on her stomach. Earhart expressed a discomfort with this role, feeling both frivolous and useless on this flight (not being allowed to fly herself) and receiving undue attention because of her sex. Anne Lindbergh, strangely, even seemed to promote the idea of her uselessness on flights, maintaining a certain decorum or even secrecy about her role on the plane. Anne wrote to her mother about her irritation with being interviewed on this subject: "I had been careful to evade her questions—I always evade them with anyone: on women's flying . . . and what does one do in the plane (I always answer vaguely, 'I enjoy flying very much.')."[40] In her response, it is as if she is truly trying to recreate the domestic sphere, as a distinctly private realm, in the air.

Women popularized aerial photography as a kind of family hobby during the interwar years; however, just as women were excluded from aviation during the wars, they were excluded from the realm of warfare aerial photography (see fig. 13). This exclusion would have drastic consequences for women in the field of cartography, as most significant developments in photogrammetry occurred during these times. Women

13. General Goddard, holding his 60-inch telephoto lens, makes explicit the connection between masculine identity and aerial photography. Courtesy of U.S. Air Force.

became involved, as technicians, in developing aerial films and maintaining cameras, though they were not allowed to fly. In 1944, Sara E. Johnson, an RCAF Photo School student, explained that "most of the men were gone now from the Photo section, and they had gone overseas, so that even if we didn't exactly serve that men could fly, we certainly served so that they could go overseas." She complained: "We would have liked to go overseas too, but there was little chance of that. They were afraid we might get hurt." Johnson's friend, Smoky Brinton, hoped to take aerial photographs herself, but her hopes were dashed: "Smoky had taken aerial photography as part of her photo course, but by the time we came along there were so many photo students I guess they couldn't get enough flying time, and that was scratched. . . . Our job was keeping the cameras running, and processing the film etc."[41] Women's job in keeping the cameras running reflected their subservient position to men during the war, indeed serving so that men might fly.

Sara Johnson's autobiography, *To Spread Their Wings,* provides a unique perspective of the lives of women Photo School students in Canada during the war. Johnson describes in detail her duties:

> Once we started working at the Photo Section and got settled in, we found that most of our work was concerned with the F24s the navigation students used for their bombing exercises, when they took pictures instead of dropping bombs, using the same release to take pictures as they would use for dropping bombs. It was our job to keep the cameras in running order, install them in the A/C [aircraft] . . . receive the film magazines from the students and take the cameras out of the planes. We developed the films, annotated them and printed them and took them . . . to the instructors. There was a small black cross in the centre of the ground glass in the camera and it represented where the bombs would have hit.

Though Johnson was sometimes required to "go along and watch the camera" during training flights over Canada, she never took aerial photos herself. She wrote of one experience flying: "F/S Melrose was going too and in fact he was the person who actually crawled into the nose of the plane and shot the pictures. . . . I sat in the middle of the plane until it was time to take the pictures, and then . . . I went back and crouched beside the camera."[42]

14. Melba and Marti Sentes, friends of Sara Johnson, were allowed to install
aerial cameras but not to take photographs. Courtesy of Sara E. Johnson.

Johnson's experience was not atypical during the war, where women often carried and maintained the camera but went back and crouched beside the camera when it was time for men to take the pictures; a clear division of labor emerged surrounding aerial photography (see fig. 14). It was obviously not that women were considered incapable of operating aerial cameras, since the cameras were often left in their sole care. Instead, it was that the war was considered a man's world, as women were excluded from combat duty. The camera no longer belonged in the hands of a pretty woman taking photographs on vacation; the camera was part of the war machine. The end of World War II only solidified this division in labor, as women were excluded from the commercial aviation industry altogether; there were simply too many capable male pilots returning from the war in search of work.

In Canada, where aerial photography became an enormous industry, women's exclusion from air combat would further lead to the male domination of this field. Johnson, for instance, temporarily worked in the Photo Center after the war, but she was discharged within the year: "Most of us worked in the main printing room. The Air Force was mapping Canada, taking the pictures on film that made 8" x 10" prints. The prints were used to make mosaics to make maps. There were a lot of new men around." When Johnson was unexpectedly and suddenly discharged, she responded: "I simply couldn't speak. . . . Probably I had been unrealistic, but it had simply never occurred to me that I would be discharged from the Air Force. It was one of the worst moments I have ever had." Johnson's autobiography, written fifty years after the war, describes her wartime experiences as the best moments of her life. Apparently, Johnson married and settled into a domestic life, like so many other postwar women who were pushed out of jobs after the men returned from war. Strangely, Johnson never mentions her married life at all, which critics have noted is a great lack in the autobiographies of other female aviators as well. Only once does she mention that she married, in the context of remembering her friend in Photo School: "The last time I heard of him was on my honeymoon in Banff."[43] Marriage, for these women in the aviation industry, appears as a footnote, a place to which they resigned themselves when they were forced to leave their professions. Their sense of adventure and excitement, however, was in the air— just as it was for men.

Thus, Amelia Earhart represents a moment in aviation history when women were allocated a brief, though complex, day in the sun as flyers. When women were suddenly pulled from this realm, during the wars, and not allowed back in, their sense of despair was palpable. Many women expressed a desire to "be men" and thus escape their expulsion from aviation. Anne Lindbergh constantly writes of her own inadequacy as a woman, discussing "how small it made me feel" to think of her gender. She writes, "Oh, I wish I belonged to that world of action and non-introspection—that superb, objective, vigorous world of [Charles Lindbergh's]." Anne Lindbergh's fascination with crossing gender lines is also reflected in her obsession with Virginia Woolf's *Orlando,* which she picked up shortly after meeting Charles Lindbergh, after not reading for months. Anne wrote to her sister: "You merged into it, so that when you walk out of it you still have bits of it sticking to you. You live through a thin veil of it for a while. . . . Orlando himself—herself—delights me."[44] Lindbergh's interest in the transgendered Orlando demonstrates how women like herself and Earhart felt caught between genders in their desire to fly. It could be said, in fact, that flying produced a liberating, though sometimes masochistic or confusing, gender instability in women.

This very instability, at the time, was constantly captured in the work of Virginia Woolf—as she struggled with finding a new perspective for women by exploring the aerial vantage point alongside the parameters of gender roles. However, just as madness is equated with "losing a sense of proportion" in *Mrs. Dalloway,* so the aerial perspective threatened to destabilize women's sense of proportion by forcing them to question their own gender stereotypes. Though this questioning could be viewed as potentially liberating, the punishment for overstepping their own bounds (madness) always loomed in the distance.

THE MASTER PLAN

Pilot autobiographies of both men and women are strikingly similar in their descriptions of the first experience of flying: a sense of disproportion, defamiliarization, and alienation, combined with a feeling of potency. Lowell Thomas's wife claimed, "I feel like a giantess flying over pygmy villages."[45] And Beryl Markham, the first woman to fly across the Atlantic from east to west, writes that flying "is at times unreal to the point where the existence of other people seems not even a reasonable probability. . . . The earth is no more your planet than is a distant star— a star shining; the plane is your planet and you are its sole inhabitant."[46] Lowell Thomas, in his first flight (over the pyramids) wrote, "When we got up a few thousand feet, I had the weird sensation that I was a spectator from another planet, astride a flying meteor, viewing the earth through a vista of a thousand years."[47] Amelia Earhart wrote that, from the air, "even mountains grow humble."[48] The French war pilot Antoine de Saint-Exupéry succinctly described the pilot's world: "I tower above a great sparkling pane, the great pane of my cockpit. Below are men— protozoa on a microscopic slide. . . . I am an icy scientist, and for me their war is a laboratory experiment."[49] Men appear to feel exhilarated, rather than threatened, by this experience of power; for women to occupy the space of an other-planetary bomber pilot, however, was to become a "giantess"—or to be disproportionate for her sex.

Beryl Markham's autobiography, *West with the Night,* is particularly interesting for the way in which it reads as a sort of conversion narrative in which the male space of the pilot comes to be understood and occupied by a woman. Markham describes her first experience of being rudely awakened by the sound of an airplane propeller in her village home in Africa: "More people had driven out from the town, compelled by the new romance of a roaring propeller—a sound that was, for me, like a white light prying through closed eyes, disturbing slumber I did not want disturbed. It was the slumber of contentment—contentment with a rudimentary, a worn scheme of life. . . . I had curiosity, but there was resentment with it." Tom Black, who flew that plane, described to Markham the feeling of power associated with flying: "'When you fly . . . you get a feeling of possession that you couldn't have if you owned all of Africa. You feel that everything you see belongs to you—all the pieces are put together, and the whole is yours. . . . It makes you feel bigger than you are.'" Initially, Markham is resistant to this perspective, claiming, "It seemed such a far away step from the warmth and the flow of life and the rhythm of life flowing with it. It was too much outside of the things one knew—to like, or even to believe." But she is seemingly seduced by her first flight with Black: "We swung over the hills and over the town and back again, and I saw how a man can be master of a craft and how a craft can be master of an element. I saw the alchemy of perspective reduce my world, and all my other life, to grains in a cup."[50] This reduction, in turn, allows her to feel the power of a giantess but with the ambivalence of a woman who has been strangely separated from the "warmth of life."

Anne Lindbergh, perhaps the most eloquent in her experience of flight, describes a similar attraction, which she then appears to displace onto her husband, Charles. In her first flight, she writes: "I had a *real* and intense *consciousness* of flying. I was overjoyed. . . . We were high above fields, and there far, far below was a small shadow as of a great bird *tearing* along the neatly marked-off field. . . . That 'bird'—it was *us*. We were over the city now; it looked like a doll's model." She continues: "No wonder he has a disregard for death—and life. This is both. . . . It was a complete and intense experience. I will not be happy until it happens again." Anne, after flying, becomes engaged in an intense infatuation with the sport, which also resonates with feelings of inappropriateness for her gender. Comparing herself to Phaëthon, she wrote, "We seemed so impertinent—as though, like that boy who tried to drive the sun, we had gone out of our bounds, arrogantly had gone into a sphere not ours." At times, she describes herself close to madness because of this new perspective: "The airplanes . . . and these long evenings will really permanently upset my equilibrium!" Anne's feeling of flying, at the same time, is also what seems to provide her with a sense of strength and liberation. After her first determined effort to "be conscious" while flying, Anne describes a sense of having created her own experience. "It was an intense pleasure," she writes, "so *real, so real*! Because I had it alone, and it was my own responsibility. . . . I *made* the effort to get it and it was *mine—my utterly possessed experience*." This feeling of making or creating her experience also parallels a sense of owning or controlling the world itself: "The country was shining and beautiful and spread out suddenly and simply like that under our feet, and I felt, How simple! *This* is the way the valley *really* looks, really is. I see it all now—how clear, how simple! It's such a little place, such a little world, and how neatly and plainly laid out. I can see it all. And I felt like God."[51] If the aerial camera offered a disproportionate sense of power, this very impersonal and detached view could also offer comfort. Booker Prize–winning novelist Graham Swift, who once worked as an aerial archeologist, claimed that the camera offered "a kind of comfort that every random, crazy thing that gets done should be monitored by some all-seeing, unfeeling, inhuman eye."[52]

Women's feeling of "impertinence" in the face of flying corresponds to their feelings of ambivalence about the "unfeeling, inhuman eye" of the plane. During the wars, this perspective would suddenly make sense—rid of women's ambivalence—as the aerial perspective provided the distance necessary for an indifference to death. War provided new artistic possibilities for the aerial photographer: quick changes in terrain from dropped bombs, mobilized troops, and military encampments could easily be captured with the photographic eye. Walter Benjamin has claimed that a fascist society "expects war to supply the artistic gratification of a sense perception that has been changed by technology."[53] The uses of photography that were developed during wartime were later applied to arresting, capturing, and ordering urban life. The camera, it seemed, could arrest the movement of disorganized traffic as easily as mobile armies, squatter settlements, and enemy encampments. The shift to urban aerial photography, indeed, was often read as essential to urban development. Melville Branch suggested, "An aerial photographic view of the city as a whole is essential for comprehensive study and understanding of the urban organism and the formulation of realistic programs for directing its growth and development."[54]

Branch, in *City Planning and Aerial Information,* pointed out the benefits of aerial photography for urban areas: "For the first time, urban places may be counted and classified." Like Saint-Exupéry's declaration ("I am an icy scientist, and for me their war is a laboratory experiment"), so the city became an object of scientific study and experimentation. Branch suggested that some of the uses for aerial photography included "slum clearance," "land development of all sorts," and "railroad extension and consolidation." Comprehensive zoning ordinances, known as "master plans," would come to dominate city-planning methods. Branch explained that the master plan "presumes that the purposes, needs, and situation of an organism as complicated as a sizable city can be conceived, projected, foreseen, and predetermined."[55]

But in order for the city to be treated as a holistic organism, it had to be seen as such—thus, the necessity of the aerial perspective. Aerial photography played into the belief that "the world could be controlled and rationally ordered if we could only picture and represent it rightly."[56]

Suddenly, it seemed, the airplane could keep pace with urban development even if the grounded individual could not. Richard Muir, professor of urban geography, explained, "The literal 'man in the street' finds it almost impossible to form an impression of the structure and layout of the town. . . . From heights of a thousand or several thousand feet, the town plan begins to emerge in a map-like form and the story of its development can then be explored."[57] Aerial photography, in many ways, enabled the emergence of city planning.[58] But in so doing, it removed power from the people—and reduced it to an alien form. Increasingly, there tended to emerge a "temptation to believe that what is seen from above has more reality than what faces one on the ground."[59] If "reality" was increasingly inaccessible from the ground because of urban congestion, the demand for the view from above would provide a way to access and control the reality of urban existence. But what aerial photography gave to the viewer it also took away—the city could only be understood at an artificial distance. Graham Swift wrote in *Out of this World*: "When did it happen? That imperceptible inversion. As if the camera no longer recorded but conferred reality. As if the world were the lost property of the camera. As if the world wanted to be claimed and possessed by the camera."[60] The airplane camera saw what those on the ground could not see; it captured the mobile environment of the metropolis and sterilized it and studied it. The camera, it seemed, would be granted the lost agency of the subject; the vision of the camera would replace human vision. It would be a view, however, that was not only distinctly male, but also informed by the "photograph then bomb" vantage point of the wartime pilot.

The derealized perspective of the aerial war photograph ultimately came to infect the vision of the city. What had at first eliminated blood and bodies in warfare could later literally clean up the city for the viewing public. It was, in many ways, the perfect instrument for the rational organization of disorganized space. Lowell Thomas described the way in which the aerial photograph naturally led to a city plan: "There is no doubt that the best way to see a city for the first time is from the window of an airplane. Instead of coming in through a lot of dirty railway yards and uninteresting factory and poorer residential sections, you get a perfect panoramic view, a view that once and for all puts a plan of the city in your mind's eye."[61] And the *Guardian* commented on the function of aerial picturebooks: "Aerial photography has become the supreme tourist board way to sell the idea of a country. No dirt, no poverty, no people to be seen; no waste, tensions or ugliness."[62] While the camera cleaned up the image of the city for public consumption, it also served to eliminate slum areas, which the interpreter was trained to recognize from the air, for the sake of future development.

France passed a law, after the Versailles Treaty, requiring that all cities above a certain size be resurveyed within three years. Because this demand would have been physically impossible by ground survey, the law essentially mandated aerial surveys of urban areas. One aerial company alone surveyed more than two hundred cities according to this mandate.[63] In the United States, the Harvard City Planning Studies series declared, "The easiest way to keep track of one's municipal neighbor, at least in many physical-spatial respects, is to periodically 'look over the back fence.'"[64] The aerial photograph as a method of surveillance was also promoted for the revenue it could bring in: "In six Connecticut towns the use of air photographs resulted in the discovery of 1,237 residences, 12,534 garages and barns and 13,866 lots which were escaping taxation." Aerial photography became a mandated method of surveillance and development that reached beyond the boundaries of the city itself to control outlying areas. Aerial photography, one proponent suggested, was "useful for the study of the nature, existing pattern, and utilization of outlying land within the urban region; it is also involved in any evaluation of its probable or most suitable future use."[65] The suitability of land tenure in peripheral areas would become further controlled as the power of the city began to stretch out as far as the aerial camera could see.

Because of the continued experimentation of the military in urban-mapping projects after the war, as well as the commercialization of war equipment and personnel, the line between military control and urban planning became slippery. For instance, one of the early assignments of General George W. Goddard in the United States after the war was to photograph coal-mining strike riots in West Virginia. He writes: "Our mission was to photograph roving mobs of miners armed with shotguns. Everything looked very warlike." The purpose of this mapping assignment was to "locate the rioters so that the military ground forces could take them into custody." Goddard also mapped the Philippines from

15. General Goddard explains the principles of aerial photography in the Philippines. Photograph from the National Anthropological Archives, Smithsonian Institution.

1927 to 1929, during U.S. military rule of that territory (see fig. 15). Another covert mapping mission involved the photographing of the Colorado River to monitor Mexico's use of water. Goddard explained that the major challenge of this expedition was to maintain a high enough altitude to be "well out of the range of any rocks the Mexicans might throw at us." Goddard also surveyed the Tennessee River Basin for the development of the TVA dam project, the largest dam project in the nation at the time, for the supply of power to urban areas. One of Goddard's urban-mapping projects was also captured in a newsreel entitled *Army Aerial Photographers Map Los Angeles from the Air.*[66]

But perhaps the most blatantly militaristic urban aerial project occurred when Goddard developed the idea of dropping bombs over U.S. urban areas at night to create a kind of aerial "flash" photography (see fig. 16). In his memoirs, he describes repeated scenes of urban panic created by these bombing "raids" during a time of peace. Of one scenario,

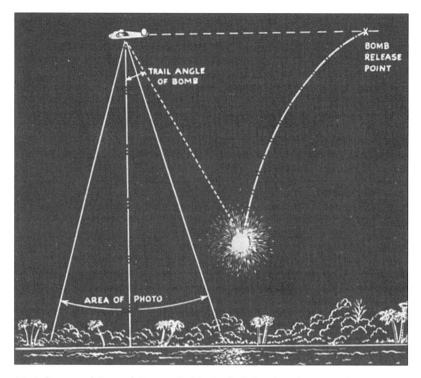

16. A diagram of General George Goddard's flash-bomb technique, whose use created panic in peacetime cities. Courtesy of U.S. Air Force.

he writes: "When the bomb went off that late at night, it raised havoc with the unsuspecting populace. . . . Telephones were out for several hours and there was a near panic in the city." But his colleague, "who was beside himself with joy" about the photos, responded to Goddard's dismay about the panic, "'What's the matter with you, George, it's the picture that counts, man, the picture!'" In another incident, a bomb that was dropped over a "new subdivision" in Dayton, Ohio, failed to detonate. Goddard explained the danger: "In our tests there was always the problem of a dud, a bomb that failed to explode. This could present an extremely hazardous situation because some unknowing person might pick it up by mistake and the results could be instantaneous, loud, and not redeemable." In a panic to find the missing Dayton bomb, Goddard and his colleagues finally discovered it stacked in an elderly man's shed. After Goddard dropped a bomb over Chicago, the *Chicago Tribune* headline read, "Gangsters Bomb Tribune Tower." According to Goddard, nobody bothered to "enlighten" the public about what had actually occurred.[67]

Goddard, who is said to have pioneered high-altitude photography, worked on reducing the amount of time it took to get the photograph into the hands of the military strategists. Film would be dropped from a plane to a ground crew stationed below, who would work in laboratory field units to get the photos developed. The photographer may not have even known how to read the photographs himself, which had to be viewed and translated through stereoscopic equipment for three-dimensional attributes. In the field of cartography, a complex breakdown would occur in the division of labor. The photographer—given an angle, a compass direction, and the area of landscape to be photographed—would turn the photograph over to a developer, who turned it over to a stereoscopic interpreter, who ultimately gave its data to a cartographer. If the photo was used for cartographic purposes, it would have to be pasted into a photo "mosaic" and traced by hand; once the necessary lines were traced from the photograph, it was often filed away and forgotten. Patrick McHaffie explains: "As the cartographic production process has been progressively fragmented and rationalized, the cartographic laborer has become increasingly alienated from the product of his or her labor: cartographic information."[68] The only perspective from which the map could be truly read as a holistic object was that of the consumer.

Fredric Jameson wrote, "The alienated city is above all a space in which people are unable to map (in their minds) either their own positions or the urban totality in which they find themselves."[69] The urban "map"—or the aerial perspective—provided its consumers with a sense of totality or placement. Women flyers, excluded from the strange realm where men sometimes "took pictures instead of dropping bombs," would also be excluded from the realm of postwar urban aerial mapping. The idea of ordering the city through the aerial perspective would become a distinctly masculine enterprise, with historic links to warfare, covert actions, and postwar military enterprise.

PHOTOGRAPHIC LOSS

Walter Benjamin, in "The Work of Art in the Age of Mechanical Reproduction," claimed that the photograph destroyed the aura of the landscape. Benjamin describes the experience of the aura in the natural world: "If, while resting on a summer afternoon, you follow with your eyes a mountain range on the horizon or a branch which casts its shadow over you, you experience the aura of those mountains, of that branch."[70] When the camera captured the city from the air, it was this experiential element that was lost.[71] The aerial photograph substitutes grounded perspective for that of the master plan, destroying the legitimacy of the human perspective, which can no longer confer a sense of reality. Sigmund Freud described that uncanny moment when "the foreign self is substituted for his own—in other words, by doubling, dividing and interchanging the self"—the double, then, is invented for the sake of "preservation against extinction."[72] The aerial photograph preserves the urban area from extinction, but only by substituting the "master plan" or "grand narrative" for situated knowledges. The photograph arrests time, stops wars, confers power and order—but only at the cost of human feeling. The Harvard City Planning Studies guide promoted the aerial photograph for educating children: "Larger scale photographs are effective in the classroom as a means of depicting the city as a reality of many parts, and photo assemblies as an instrument in promoting an awareness and comprehension of the organism as a whole."[73] The holistic vision of a city as simultaneously "many parts" would provide a new way of thinking about urban areas. Rather than class-divided pockets of territory, the urban area would suddenly become a unit that could be perceived and manipulated as a whole. The frenetic desire to be captured by the photograph would outweigh what it also seemed to steal—experience—because somehow only the photograph seemed able to confer reality and permanence. As General Goddard's friend exclaimed, "It's the picture that counts, man, the picture!"

Benjamin described this desire: "Every day the urge grows stronger to get hold of an object at very close range by way of its likeness, its reproduction . . . to pry an object from its shell, to destroy its aura."[74] The desire to get hold of the city for purposes of control would prove insatiable. As aerial photography grew in popularity, city-planning committees began to rely upon it more and more in making their decisions. In the process, the agency of those people living in the photographs began to be eroded. In 1965, Aerofilms wrote: "We are frequently approached, too, before things get to the planning appeal stage, to photograph a large number of prospective sites for a particular project. . . . This is cheaper and more desirable than the local visits of surveyors and inspectors, whose mere presence causes speculation and rumor in the locality."[75] Commercial aerial photography became a way to evade the speculation and rumor of the locals who might want to be included in the planning process. Similarly, wartime aerial photography allowed one side to outmaneuver the movements of the other, to gain perspective of the strategy, and to see the enemy as an object to overcome. The dependency upon the aerial perspective for urban control would, however, become a dependency upon the double of the self—the alien perspective—to see. The aerial photograph would reify abstract possibilities of control, which would become culturally gendered as male, even though it was no longer human. "The pilot cyborg is an enhanced human, a man-plus," suggests *The Cyborg Handbook*.[76] The man-plus of the camera-photographer with the power to drop bombs would overtake the city itself as an object of its possession. In so doing, however, a loss in spatial proportion would occur that would seem impossible to resurrect—except as a double. Although people might strive for connection and lament its loss, the possibility for community would be increasingly removed to the impossible space of the sky.

Beryl Markham, in her transatlantic flight, describes the sensation of flying: "Being alone in an aeroplane for even so short a time as a night and a day, irrevocably alone, with nothing to observe but your instruments and your own hands . . . such an experience can be as startling as the first awareness of a stranger walking by your side at night. You are the stranger."[77] Not only does Markham see herself as a cyborglike pair of "instruments and hands," she also sees herself as a particular kind of stranger: a frightening one. The "startling" experience she describes of a stranger suddenly walking by her side at night contains obvious overtones of male assault. But, in an interesting twist, Markham realizes that *she* is the assailant. In this passage, she expresses the startling realization that she not only has become cyborg, but also has become man. And in this role, she realizes, she no longer recognizes herself.

Markham's experience of being taken over—or assaulted—by a stranger at night could be said to represent the fears that women in the aviation industry experienced as a whole. Not only did women fear their own gender as a liability, and thus identify with male pilots, but also they saw themselves as being pushed out of the industry by the men. As woman's perspective of flight was interrupted by war, colonized by the propaganda of the heroic flying ace, then further erased by postwar militaristic mapping pursuits, it must have seemed that men were taking over the roles that women once filled, completely occupying women's space in the air. The man holding the aerial camera, which ordered cities and plotted wars, ultimately undid the vision of these women, who were instead arrested in the camera's gaze.

CHAPTER 4

Air Control, Zerzura, and the Mapping of the Libyan Desert

In Chicago's lakefront park, on a thin strip of grass between Soldier Field Stadium and the jogging path, there is an unobtrusive statue with an Italian inscription. It is the only remaining structure from the Chicago World's Fair of 1933, a column removed from the ruins of a Roman temple. The inscription on the column reads:

FASCIST ITALY WITH THE SPONSORSHIP OF BENITO MUSSOLINI
PRESENTS TO CHICAGO
AS A SYMBOL AND MEMORIAL IN HONOR
OF THE ATLANTIC SQUADRON
LED BY BALBO
WHICH WITH ROMAN DARING FLEW ACROSS THE OCEAN
IN THE ELEVENTH YEAR
OF THE FASCIST ERA

This column, sent to the mayor of Chicago by Benito Mussolini, commemorates Italo Balbo, who led a transatlantic flight of twenty-five Italian seaplanes to Lake Michigan for the opening of the "Century of Progress" World's Fair. The motto of the fair was "Science finds, industry applies, man conforms." Balbo's flight was one of the most highly publicized and photographed of the transatlantic flights, creating an aerial spectacle that "showed the Italian flag, publicized fascism, displayed the Aeronautica's technical prowess, and demonstrated the excellence of the Italian aircraft industry to prospective customers."[1] Balbo, the heir apparent of Mussolini, was greeted with the fascist salute upon his arrival in Chicago. He was honored with a street named after him in Chicago, a New York ticker-tape parade, and lunch with President Franklin D. Roosevelt.

In the 1920s and 1930s, a fixation with aviation and a kind of technological determinism emerged, often under the banner of fascism. Fascism, at the time, could be German, Italian, British, or American. It crossed national boundaries and was often united in the purpose of controlling the colonies. The aviators themselves—including such figures as Charles Lindbergh, T. E. Lawrence, Italo Balbo, and László Almásy—tended to share an affinity with fascism, as well as a technocratic belief in the airplane as a tool for both controlling and exploring the colonies. Almásy's family was part of an aristocratic group of pan-national Nazi sympathizers. Lindbergh also maintained a flirtation with fascism that would mar his career until his death. T. E. Lawrence (Lawrence of Arabia) was part of Britain's "Cliveden set," which supported National Socialist Germany and Fascist Italy as a bulwark against Soviet Communism. This aristocratic association of Nazi sympathizers also included such prominent figures as Lord and Lady Astor, Geoffrey Dawson, editor of the *Times,* and Lionel Curtis. Henry Williamson suggested that Lawrence should become the dictator of England. He explained, "The new age must begin . . . Hitler and Lawrence must meet." On the morning of May 13, 1935, Lawrence was killed in a motorcycle accident while driving back from the post office after sending a telegram to Williamson, agreeing on a time to meet Hitler.

T. E. Lawrence, one of Balbo's admirers, popularized the use of aerial reconnaissance against desert guerrillas in the Middle East. His successful desert campaigns—with airplanes and armored cars—led Britain to adopt his strategies for "imperial policing" in the colonies between 1919 and 1939. In order to "deal with restive populations and disorders of all sorts in its empire," Britain perfected the method of "air control" in the Middle East and Africa.[2] The operations involved in air control strategies were:

1. Develop a statement of what was expected from a "target tribe."
2. Drop leaflets or otherwise inform the population of these expectations.
3. Bomb the village(s) if they refused to comply.

Captain David Parsons of the U.S. Air Force wrote, "The bombings, interspersed with deliveries of propaganda literature, would slowly intensify until the recipients sued for peace on terms acceptable to the government. According to RAF policy, the stated political objective of air control was 'to bring about a change in the temper or intention of the person or body of persons who are disturbing the peace. . . . In other words, we want a change of heart.' "[3] This idea of control without occupation would prove to be more economically viable than colonization. In 1997, Colonel Kenneth J. Alnwick, also of the U.S. Air Force, applauded the effectiveness of air control, explaining that "the harassed tribe recognized the reasonableness of British demands and the benign nature of British colonial administration."[4] In order to inform the target tribe of "expectations," however, the tribe had first to be found. It was discovered, at this time, that aerial photography was very useful in locating the mobile desert guerrillas in hidden or remote oases. Lawrence, after the war, joined the School of Aerial Photography at Farnborough in England in order to perfect this strategy.[5]

In "The Art of Aerial Warfare," Balbo explained that the massive air formations of the Italian Aeronautica taught the values of the military drill: discipline, organization, and group mentality. Claudio Segré described Balbo as "one who made a career out of organizing and leading masses, whether Blackshirts or colonists or aviators."[6] In the deserts of Libya, Balbo introduced the art of aerial warfare against the local populations. Following the World's Fair of 1933, Mussolini appointed Balbo to govern the colony of Libya, referred to popularly as "Italy's fourth shore." Balbo's rule would not be as violent as that of his predecessor, Rodolfo Graziani—but due to their combined efforts to pacify the colony, half of Libya's population would be killed, starved, or forced into exile between 1912 and 1943.[7] In 1933, the Italian Army Health Department chairman, Dr. Todesky, wrote: "From May 1930 to September 1930 more than 80,000 Libyans were forced to leave their land and live in concentration camps. . . . By the end of 1930 all Libyans who lived in tents were forced to go and live in the camps. 55% of the Libyans died in the camps."[8] Those who resisted Italy's rule—primarily the Senussi—would hide out in caves and desert oases, making raids on Italian arms depots and receiving smuggled weapons from neighboring Egypt.

In 1931, the last Senussi stronghold, Kufra, fell to the Italians, and the leader of the resistance, Omar Mukhtar, was hung before a crowd of twenty thousand. Khaled Mattawa, an Arab American poet from Libya, described this time in Libya's history in a poem called "General Italo Balbo: Tripoli, 1937."

> It's safe to walk
> the streets now, the rebels long subdued
> by Graziani. In the square you stroll
> he strung up hundreds, once leaving
> five dangling for a week until a film crew
> (experimenting with color) arrived from Rome.
> This is not your method. The few
> you catch now are shot far away,
> two bullets to the head, unmarked graves.

Balbo organized a massive immigration campaign of Italians to Libya to develop the interior and the Libyan Desert. Mattawa comments on Balbo's boosterism regarding the potential of desert lands for settlement:

> And why
> would they not believe you General . . .
> the photogenic ex-veteran?
> They will believe
> you "Il Padre D'Aeronautica" who crossed
> the Atlantic leading a fleet of hydroplanes,
> star of the Chicago World's Fair. "Balbo,
> Balbo," New York greeted you with downpours
> of confetti in a Broadway ticker tape parade.
> Roosevelt shook your hand firmly two days later,
> poured your coffee, another medal on your breast.[9]

Balbo's campaign to colonize the Libyan interior was inspired, in part, by a desire to manifest sovereignty in this desert frontier. The Libyan Desert, the northernmost part of the Sahara, remained one of the last unexplored blank spots on earth, except for the polar regions, in the early twentieth century. Crossing the boundaries of Egypt, Sudan, Chad, and Libya, the Libyan Desert was controlled in the 1930s by the British, Italians, and French (see fig. 17). Prior to the advent of the automobile, the Libyan Desert had been explored only by camels, which could not travel far from known water sources. Sir John Gardner Wilkinson traveled in the desert by camel in 1838 and heard about an oasis called

"Zerzura" from the inhabitants of the Dakhla Oasis in Egypt. Harding King, seventy years later, made a series of camel journeys into the Libyan Desert and also recorded legends of the famous Zerzura oasis, a mythical place with hidden treasures. He also told of its "black raiders" occasionally attacking other Egyptian oases.[10]

This dual legend of black raiders/mythical oasis would come to dominate colonial descriptions of the Libyan Desert. The mythical raiders would come to be linked with the resistance to Italian rule by the Libyan Mujahedeen (freedom fighters). The impartial pursuit of geographic knowledge, therefore, became sometimes indistinguishable from the violent conquest of Libya. The British, German, and Italian policing of the Libyan Desert would, in turn, revolutionize the way in which cartographic enterprises proceeded in desert colonies. Segré explained that "in place of the usual camel corps (*meharisti*), Balbo devised a mixed unit of aircraft, camel corps, and motorized riflemen and machine gunners."[11] These methods, in turn, would be adopted by the explorers of the Libyan Desert. British explorer Ralph Bagnold commented that "the Sudan officials had recommended mounting machine guns on our cars."[12] Though Bagnold does not say if he adopted this strategy, the combination of auto-

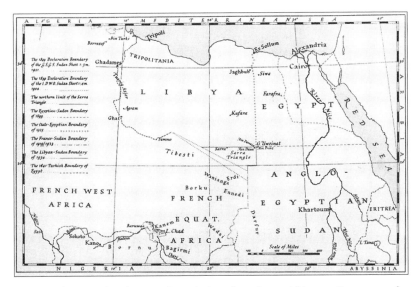

17. Map showing the changing boundaries of northeast Africa, as European colonial powers raced to claim territory (1935). Photograph from the Royal Geographical Society, London.

mobile and airplane would become the standard way of exploring the Libyan Desert. Explorers would drive as far as possible, then take an airplane to photograph regions that were inaccessible by automobile.

László Almásy, an explorer of the Libyan Desert in the 1930s, claimed that "if one or several rain oases with at least intermittent springs existed in the Gilf Kebir, many dark problems of African history could easily be explained, from the campaign of King Cambyses down to the Senussi raids during the world war."[13] One of the "dark problems" of African history was what Ralph Bagnold called "the Senussi menace." In Libya, the Senussi repelled Italy's invasion until Italy "issued a proclamation that the tribes must either obey the Italian government or be destroyed as rebels. It was to be war with gloves off." Bagnold explained: "There was no real alternative. Once having made up their minds to take possession of the country, it was impossible to leave independent marauding bands free to make surprise raids, to disappear into the interior and turn up again elsewhere to raid again." The pursuit of the Senussi led to desert exploration by motor car in Libya, facilitating more intensive exploration into the interior.

The attitude of the explorers toward the pacification of Libya was ambivalent, at best, but mainly sympathetic to the Italians. In the 1930s, Kufra, the last oasis to fall to the Italians, was also the starting point for expeditions into the Libyan Desert (see fig. 20). Kufra became a kind of oasis of transnationalism as explorers from Britain, Hungary, and Germany mingled with the Italians and resident Senussi. A companion of Almásy, H.W.G.J. Penderel, wrote of Kufra: "On our arrival at Kufara [sic] we were received by the Military Governor and the officers of the garrison. Their hospitality and the help they gave us were unlimited and left us greatly in their debt."[14] Similarly, Dorothy Clayton-East-Clayton (also known as Dorothy Clayton), the only female explorer of the Libyan Desert, wrote in 1932:

> At Kufra, the Italians gave us a great welcome. I had previously on my way out by air made acquaintance with the hospitality of an Italian Air Force mess and knew that we should find friends, but I was almost overcome with their welcome. I should also like to place on record that both in Italy and in North Africa it would be impossible to find greater helpfulness, hospitality and good company than with

the Italian Air Force. During this and another visit to Kufra a week or two later I made a tour of several of the villages of the oasis. The Arab population has somewhat declined as the result of the occupation, the uncompromising fanaticism of the Sennusiya making it difficult for them to live at close quarters with the Italians. Many wandered out into the desert to direct starvation and thirst after their resistance had been broken in 1931. Others crossed the border into Egyptian territory.[15]

Armed bands of Senussi were "hunted down with aircraft and motor columns, and dealt with ruthlessly with bombs and machine-gun fire wherever found."[16] Because it was believed that these raiders hid out in desert oases, a fixation with discovering these oases started to overlap with a militaristic need for finding rebels.

The aerial camera, it was believed, would ultimately become the most effective way of maintaining control in remote colonies. During the 1930s, Almásy and a group of explorers extensively used the camera, airplane, and automobile to map the remote regions of the Libyan Desert. Ralph Bagnold successfully crossed the "Sand Sea" for the first time in 1930, covering more than four thousand miles in Model T Fords (see fig. 18). Patrick Clayton of the Egyptian Desert Survey worked on

18. Extricating one of Ralph Bagnold's Model T's from the soft Saharan sand. Explorers spent much of their time digging out cars in the desert sand. Photograph from the Royal Geographical Society, London.

19. P. A. Clayton, surveying the Libyan Desert with his assistant, who spent much of his time protecting Clayton and his equipment from the hot desert sun. Photograph from the Royal Geographical Society, London.

systematic mapping of the Libyan Desert by airplane and automobile, following in Bagnold's path (see fig. 19). The president of the Royal Geographical Society claimed that "the opening up of the Libyan Desert . . . is filling what was a most humiliating blank on the map. I think that must be one of the reasons for the lively interest in that area. Another reason perhaps why attention has been paid to it is that it has an air of mystery about it."[17] The Libyan Desert gradually became so accessible that one explorer declared, "The sand sea has become as open to cars as the ocean to ships."[18] In the face of this methodical occupation, explorers raced to find the last remaining mysteries of the desert.

In 1932, Zerzura was discovered—via aerial camera—by László Almásy, who ultimately became the main character of Michael Ondaatje's *English Patient.* The myth of black raiders and beautiful oases would gradually die as the Libyan Desert was pacified and the Senussi killed. The historic search for Zerzura created an idealized, though temporary, pan-national, homosocial environment in the desert, in which explorers constantly lamented the loss of the very thing that they were trying to discover. Believing their pursuit to be unmotivated by anything except the desire for pure scientific knowledge, they saw themselves as heroically rising above petty national allegiances. Instead, the search for Zerzura was seen as a sign of Anglo-European superiority, a pan-nationalism that begat a kind of brotherly pan-fascism in the desert.

By 1936, Balbo erected a huge travertine marble arch for Mussolini in the desert, brought over piece by piece from Rome. Mussolini said, at the unveiling ceremony, "Be proud to have left this sign of fascist power in the desert." Balbo hoped that, one day, roads would cross the Libyan Desert, irrigation projects would flood it, and a hundred thousand Italians would live there. The travertine arch signified this occupation campaign, and Balbo brought out fresh raw vegetables to represent the crops that could be raised in Libya. A French journalist remarked, "It was like finding roses at the North Pole," to which Balbo responded, "If an Italian has reached the North Pole, you'll find roses there."[19] As roses were symbolically brought to the North Pole or the Libyan Desert, the last two blank spaces on earth, Almásy and other explorers began to lament the lost age of exploration—and the brothers that it created. The desert was becoming settled, gardened, feminized; it was becoming lost, even as the masculine enterprise of aerial armadas, photography campaigns, and machine gun battles was demonstrating the technical mastery of fascism. Zerzura emerged in this brief moment between the aerial armada and the garden, between Balbo's war and plan of occupation, and between the two world wars. Though the pacification of Libya enabled further exploration into the desert, it also enabled the travertine arches and rose gardens that were the very signs of civilization that the explorers were trying to escape.

A PHOTOGRAPHIC FIND

Michael Ondaatje chose to write on the life of László Almásy after visiting the Royal Geographical Society in London, claiming that it "opened up a whole world of explorers, and a way of seeing the world." He explained: "What was useful in the Royal Geographical was not so much the information, as it was their manner of writing. . . . No complaints. No praise. It was just, you had to get from here to there. . . . So many kilometers. There's a waterhole here." To the *New York Express,* Ondaatje said: "I was fascinated as I read these explorers' writings. I didn't want to read books about the desert by writers; it all would be about sunsets and how sensitive they were. . . . Explorers wrote about crossing 75 km of desert and waterholes, they never talked about sunsets."[20] Ironically, Ondaatje's writing—stylistically—is far from a "so many kilometers" style. It has been praised for its intense lyricism, often shifting from poem to dream to description.

Ondaatje's Almásy is in the Libyan Desert for the purpose of making maps, but he also continually expresses his ambivalence toward maps. Almásy describes his reasons for entering the desert: "All I desired was to walk upon such an earth that had no maps"[21] Almásy repeatedly refers to the desert as a place where the effects and prejudices of civilization can be stripped from the body:

> The desert could not be claimed or owned—it was a piece of cloth carried by winds, never held down by stones, and given a hundred shifting names. . . . All of us, even those with European homes and children in the distance, wished to remove the clothing of our countries. It was a place of faith. We disappeared into the landscape. . . . Erase the family name! Erase nations! I was taught such things by the desert. (139)

Burned beyond recognition, Almásy represents the art of disappearing to an extreme. He becomes a part of the desert even in his death, as Caravaggio, his fellow refugee at the Italian villa, describes him: "It is mostly the desert now. The English garden is wearing thin. He's dying" (164). Death, in *The English Patient,* becomes the supreme testimony of the ability to disappear, escaping civilization to become a part of the desert.

In *The English Patient,* Almásy enters the Libyan Desert with "some camels a horse and a dog" (136) as a solitary explorer on foot who sometimes "joined a Bedouin caravan" in his travels (138).[22] This image of the half-Bedouin nomadic explorer who has escaped the effects of civilization reads much like the romanticized portrayals of T. E. Lawrence, such as *Lawrence of Arabia.* In reality, Lawrence was more likely to travel by plane than camel, and Almásy worked as a car salesman for Steyr, an Austrian car manufacturer. Almásy's fascination with motorization appears to have preceded his interest in the desert. According to a friend of Almásy's: "At ten years old he drove his father's one cylinder Oldsmobile, his school books were full of drawings of cars and aircraft. To the fury of his teachers, when he was at school in Graz at age fourteen, he constructed a glider, took off from a quarry and smashed it and three of his ribs." He received his first pilot's license when he was seventeen, and by age nineteen had assembled and flown his own aircraft.[23] Later, he also arranged African safaris by automobile for the Austro-Hungarian aristocracy, including Duke Antal Esterhazy, Count Szigmond Szechenyi, and Prince Ferdinand of Liechtenstein. His brother, János, organized hunting retreats at Berstein, the family estate in Hungary, for several Egyptian princes, including Prince Youssef Kamal. Almásy was at the end of the age of the gentleman-explorer, a time when many of the new heroes were engineers. He combined the fascinating qualities of an almost feudal romance with the country estate and the ability to engineer or develop Africa. The desert, then, became a way to escape into the feudal past of the princely gentleman, while the car, airplane, and camera represented ways to modernize that very desert.

It is the same confusion between modernization and idealization of the premodern that is evident both in Almásy's journals and *The English Patient*. If, on the one hand, Ondaatje valorizes the pragmatic approach of the explorers, on the other hand he nostalgically romanticizes the lost age of the gentleman-explorer: "These explorers in the 1930s were out of time. I love the idea of them checking out sand dune formations. I love historical obsessives. . . . And this wonderful, heroic era of exploration that was then ignored, while the twentieth century became more mercenary or mercantile."[24] Like Lawrence, Ondaatje combines a romanticized notion of the "heroic era of exploration" with a pragmatic sense of "you had to get from here to there."

The explorers, who believed with missionary zeal that they were bringing knowledge to the desert, had to first discount local knowledge as illegitimate. In 1818, one explorer asked the rhetorical question, "Do the Hottentots of the Cape—do the more civilized tribes of African negroes . . . know the extent of their respective countries?"[25] Similarly, in *The English Patient*, Almásy says that, after his plane crashed in the desert, he was kept alive by the Bedouin for his cartographic knowledge of their territory. Ondaatje's Almásy, like Kipling's character Kim, is noted for his photographic—or cartographic—memory. Almásy has the ability to memorize details that will quickly make sense of any situation. His usefulness to the Bedouin tribe is precisely that he "draws maps that go beyond [the Bedouin's] own boundaries," in this way extending his photographic eye to the local populations (22). He claims that the

Bedouin kept him alive only because "I have always had information like a sea within me. . . . So I knew their place before I crashed among them, knew when Alexander had traversed it in an earlier age" (18). In this way, the explorers, and Ondaatje, romanticize "going native," or stripping themselves of their national garb, yet also insist on displacing the natives with their own historic and photographic vision of the world.

Almásy brings Alexander to the Bedouin, displacing local narratives with a colonial map of the world. By grounding Zerzura in historical legend, explorers were not only locating it in an imperial history but also identifying themselves with a group of legendary forefathers such as Herodotus. These stories seemed to create a kind of homosocial bond between the Zerzura explorers, making them long not only for the other legendary male explorers but also for the feelings of excitement they had in each other's presence while recounting these stories. Almásy, in *The English Patient,* described the explorers as spending their free time "placing Arabic texts and European histories beside each other in an attempt to recognize echo" (153). Christopher Miller claims that a certain textual overlapping was common in Africanistic texts: "Texts on Africa were severely limited in number until the nineteenth century and tended to repeat each other in a sort of cannibalistic, plagiarizing intertextuality. Pliny repeats Herodotus, who repeats Homer, just as later French and English writers will copy each other and even copy the Ancients." But what Herodotus could not understand, like other Africanists, were the people who live in Africa. He described the natives of the desert as unintelligible "dwarfs," "troglodytes," "dog-eared men," or "headless" men with "eyes in their chest."[26] He continues, "They eat snakes and lizards and other reptiles and speak a language like no other, but squeak like bats."[27] These people who "squeak like bats" were the unrepresented, without texts that would echo beside the English.

In *The English Patient,* the idealization of the desert is linked to colonial intertextuality as well as a space between nations. Almásy is obsessed with Herodotus's *Histories,* which he uses as a journal, including "maps, diary entries, writings in many languages, paragraphs cut out of other books. All that is missing is his own name" (96). Hana reads *Kim* and *The Last of the Mohicans* to Almásy. *Kim* is about an orphaned Irish boy camouflaging himself as an Indian. *The Last of the Mohicans* is the story of a white boy growing up with the Mohicans, but torn between the Mohicans, Hurons, British, and French during the Seven Years' War. Ondaatje's explorers are similarly thrown together in a space of representational exile, as Almásy claims: "We were German, English, Hungarian, African. . . . Gradually we became nationless. I came to hate nations" (8). Each of these boy-adventurer-explorer narratives doubles as a narrative of hybridized identity, as well as a narrative of homoerotic associations.

If the elusive Zerzura could never be discovered, it could at least lead to a solid association of men who were able to step outside modern civilization and its demands and go native, like Kim or the last of the Mohicans. In *The English Patient,* Almásy claims: "We sailed into the past. We were young. We knew power and great finance were temporary things. We all slept with Herodotus" (142). Lawrence recounts even more explicit moments of homoeroticism in his autobiography, *Seven Pillars of Wisdom,* where he describes "friends quivering together in the yielding sand with intimate hot limbs in supreme embrace." These sexual acts, he explains, were "a sensual coefficient of the mental passion which was welding our souls and spirits in one flaming effort."[28] In *The English Patient,* Almásy is also taken to a "village of no women." He says: "Here in the desert, which had been an old sea, nothing was strapped down or permanent, everything drifted—like the shift of linen across the boy as if he were embracing or freeing himself from an ocean or his own blue afterbirth. A boy arousing himself, his genitals against the colour of fire." In this scene, "one of the men crawls forward and collects the semen" and brings it over "to the white translator of guns," Almásy, and "passes it into his hands" (23). "Sleeping with Herodotus" not only binds the men to each other, but also allows for a kind of copulation with a non-European discourse, boy natives and explorers, Arabic and English. If the desert forces men to strip off the clothes of their nationality, it also allows them to disappear, momentarily, in a homoerotic embrace.

Fantasies of masculine intertextuality also imbued the historical quest for Zerzura. Bermann describes how his interest in the Libyan Desert was "excited" by his Hungarian friend Almásy: "Since the times of the ancient Egyptians, he explained, there had been a vague but persistent rumour about fertile lands lying in the desert west of the Nile. . . . All through the Middle Ages Arab writers told about a hidden oasis [by] the name of *Zerzura*—meaning probably 'Oasis of Little Birds.'"[29] Penderel describes a similar experience in which "Almásy would quote to me the

Arab legends of the 'Book of Hidden Treasures' (Kitab el Kanuz), where
the fabled Zerzura is described as 'a white city, white as a dove.'"[30]
According to the legend, the city of Zerzura was full of riches, guarded
by a sleeping king and queen. The legend warned that the explorer must
not wake the king and queen, but only take the treasure.

    In the 1930s, the oasis of Zerzura became the object of pursuit, a
mystery often linked to what was perceived as the tangled reality of the
Bedouin mind. Bagnold wrote: "It is possible that under the mental stress
of solitude and fear his [the Bedouin's] wish-oasis might become a memory of an imagined reality which he would afterwards describe to others.
. . . But the bedouin is not given to self-analysis or to disentangling facts
from fancy, so we shall never know the truth."[31] Bermann writes of trying to extract the truth from the natives, expressing an element of frustration at their silence, as well as an aspect of coercion: "We had reasons
to expect that the natives of Kufra, especially the Tebus, had always known
of the existence of these wadis, and so we tried to make them speak."[32]
Similarly, in *The English Patient,* Michael Ondaatje describes this scene:
"Bermann and I talk to a snakelike mysterious old man in the fortress of
El Jof. . . . We talk to him all day, all night, and he gives nothing away. The
Senussi creed, their foremost doctrine, is still not to reveal the secrets of
the desert to strangers" (140).[33] Zerzura, it seemed, was always more of a
driving force than an actual place, which became linked—metaphori-
cally—with knowing or understanding the Bedouin people.

    Zerzura would come to represent the confused fantasy or last secret
of the Bedouin mind, as narratives of discovery were linked to the extri-
cation of secrets from native guides. Zerzura, of course, had never been
lost by the natives and was never a mythical oasis to the Tebus. Winona
LaDuke, an Ojibwe activist, similarly wrote of Columbus's discovery of
America: "To 'discover' implies that something is lost. Something was
lost, and it was Columbus. Unfortunately, he did not discover himself in
the process of his lostness. He went on to destroy peoples, land, and
ecosystems in his search for material wealth and riches."[34] Similarly, the
only people lost in Libya were the explorers, who relied upon native
guides to extract the mystery of Zerzura; at the same time, they were
actively involved in destroying the very people upon whom they relied.

    David Spurr, in *The Rhetoric of Empire,* has suggested that colonizers
projected—particularly onto Africa—a "radical absence in Western con-

sciousness." In this sense, the search for Zerzura reflected a doubled consciousness, a "moving simultaneously in two directions: the expansive forward movement of technological development and, along with this, the confrontation with a metaphysical nothingness."[35] As Michael Ondaatje wrote in *The English Patient,* "The English love Africa . . . a part of their brain mimics the desert perfectly" (33). What the colonizers may have been trying to extract from the native, in reality, was the native himself; to appropriate the native for themselves, or go native. The English saw the locals as the primitive side of themselves, which they both longed for and hoped to civilize. Spurr describes the way that this ambivalent attitude about Africa led to a kind of language in which "rationalized language combines with the most heightened idealization of Arab character."[36] The mapping eye, or the camera, is ambivalently joined to the eye of the explorer who longs for a world without maps, who longs for the eyes of a Bedouin. Ondaatje, writing in an Africanist vein, demonstrates both a fascination for the "rationalized language" of recording detail and a romanticized view of the desert and the Bedouin.

T. E. Lawrence, in *Seven Pillars of Wisdom,* similarly romanticized Bedouin primitivism: "The Beduin of the desert . . . had embraced with all his soul this nakedness too harsh for volunteers, for the reason, felt but inarticulate, that there he found himself indubitably free. . . . In his life he had air and winds, sun and light, open spaces and a great emptiness."[37] Bagnold claimed to have been drawn to a "seducing blankness" in the desert, claiming that, for the "scientifically minded," Zerzura offered "an expectancy of finding anything that is not yet known," such as an archaeological site or a new plant or mineral.[38] But for the "less scientifically minded," Bagnold described a "still more vague" kind of desire: "An excuse for the childish craving so many grown-ups harbour secretly to break away from civilisation, to face the elements at close quarters as did our savage ancestors, returning temporarily to their life of primitive simplicity and physical vigour; being short of water to be obliged to go unwashed; having no kit, to live in rags and sleep in the open without a bed." [39] Almásy also romanticized desert life, writing: "I love the desert. I love the endless wasteland in the trembling mirror of the *fata morgana,* the wild, ragged peaks, the dune chains similar to rigid waves of the ocean. And I love the simple, rugged life in a primitive camp in the ice-cold, star-lit night and in the hot sandstorm alike."[40] Going primitive, for these explorers, was seen as equal to encountering that metaphysical nothingness—the blank space on the map—that civilization had eliminated.

In *The English Patient,* Ondaatje's Almásy says, "There is God only in the desert . . . outside there is trade and power" (250). Jean Baudrillard also describes the desert as a metaphor for negativity, or the symbolic annihilation of civilized culture: "Deserts . . . denote the emptiness, the radical nudity that is the background to every human institution. . . . They form the mental frontier where the projects of civilization run into the ground."[41] Gilles Deleuze and Félix Guattari claim that in the desert, man becomes "the subject of a deterritorialized knowledge that links him directly to God."[42] The desert is where one finds nothingness, namelessness, God.

Zerzura was the place where Europeans could escape European civilization. They constantly expressed fear that finding the solution to the mystery would destroy its very mysteriousness. "Personally I almost hope the beautiful old legend will never quite be cleared up by a mere discovery," Bermann said to the Royal Geographical Society.[43] Similarly, Bagnold claimed that the search for Zerzura would never end:

> There is no fear that the quest will end, even though the blank spaces on the map get smaller and smaller. For Zerzura can never be identified. Many discoveries will be made in the course of the search, discoveries which will make the seekers very happy, but none will surely be Zerzura. . . . As long as any part of the world remains uninhabited, Zerzura will be there, still to be discovered. As time goes on it will become smaller, more delicate and specialised, but it will be there.

Zerzura could not be found, it was believed, because it was the last hope of the European world that something other than itself existed. Zerzura, according to Bagnold, may take "different shapes in the minds of different individuals," but it would always be out there. Symbolically, Bagnold saw Zerzura as "an idea for which we have no apt word in English"— that is, "something waiting to be discovered."[44]

This narrative of European obsession with blank spaces, oases, or lost villages in Africa is constantly repeated in Africanist texts, from *Heart of*

Darkness to *Seven Pillars of Wisdom.* For instance, the French writer Michel Vieuchange became obsessed with a place in the Western Sahara called Smara in the 1930s. Like Zerzura, it was believed to be a center for Muslim fanaticism and resistance to European rule. Vieuchange disguised himself as a Berber woman and reached the city, only to find it deserted. He wrote: "Smara, town of our illusions . . . / As ravishers we press towards thee, / But also as penitents we come. / And to the friend, or to one who questions us on the way, / we shall say, 'I know you not.' "[45] Christopher Miller, in *Blank Darkness,* sums up this ambivalent approach to Smara, or Zerzura, in Africanistic texts: "Knowledge (*science*) is light. But the miraculous appearance of the white race among the blacks here, inaugurating history and knowledge, must destroy what it 'knows.' Darkness can be known only by shedding light on it; that is, it cannot be 'known' as such. The writer who insists on detailing that kind of knowledge is working on an Africanistic ground."[46] Science and metaphysics thus enter the desert together: the camera and God, detailing and denying existence at once. The camera, it seems, is only able to photograph death.

In a 1926 documentary of Almásy's journeys, *Across Africa in an Automobile,* Almásy kills a lion, which is called the natural king of Africa. The dramatization of this event shows Almásy being persuaded by the local villagers to "free us from the king of the animals who afflicts our herds." After the lion is killed, the caption reads, "The village is freed from its oppressor." This dramatized reenactment playfully foreshadows the next image: a statue of "Lord Kitchener, Liberator of the Sudan." The entire movie, in effect, is presented as a lighthearted adventure in which the ease brought by superior European technology is celebrated and even shared with the Africans, for their own liberation. Before Almásy pulls the trigger to kill the lion, he instructs the natives to get the camera ready to record the moment. The caption reads, "Gun and camera together." Almásy may be able to construct himself as white liberator by shooting the lion, but at the same time, the object of fascination—the lion—is dead. The gun and the camera together stand irrevocably yoked, demonstrating both desire for the object and the inevitability of its death. Similarly, the explorers of the Libyan Desert traveled with cameras and machine guns, recording and eliminating the territory of the Senussi. It is in this sense that the fascinated desire for unmapped territory becomes confused, as the explorer gradually eliminates the very thing he is seeking.

Zerzura was finally discovered in 1932 by camera. Almásy, along with Robert Clayton-East-Clayton, Patrick A. Clayton, and H.W.G.J. Penderel, had taken three Ford Model Ts and a Moth airplane in search of Zerzura. The *Times* for May 6, 1932, headlined: "'Lost' Oasis in the Libyan Desert: A Photographic Find." The article states, "An enlargement of one photograph shows distinctly a white spot among the trees in the wadi, which the expedition discovered. . . . Experts are agreed that the white spot is apparently a hut. The discovery offers evidence that the wadi was recently inhabited, and it tends to confirm the belief that it is identical with the 'lost' oasis of Zerzura." On June 8, 1932, explorer R.A. Bagnold wrote: "They are very satisfied with themselves and call their wadi ZarZura. Probably there are lots more wadis. Kamel el Din says he knows all about it and it is called Wadi Malik!! . . . I saw most of their photos—they took hundreds—and saw the trees—they are genuine."[47]

But just as the "experts" were recording and interpreting Zerzura, its meaning appeared to be lost. In fact, newspaper reports continued to mention that Zerzura might never by found. In "The Lost Oasis: Across the Sand Sea," the *Times* reported on September 16, 1933:

> If the problem of Zerzura still remains unsolved, an area in which there are wadis with trees and some vegetation had been found. There may even be more than one such valley where recently none was known to exist. When all these have been visited and the Oasis of Birds has still not been located, then we shall have narrowed down even further the Zerzura problem, perhaps to a vanishing point; but until that has been done the lost oasis is still there to be found.

Earlier that year, Almásy and Clayton had entered Zerzura, which was thought to consist of three separate oases, by foot. These wadis—called Hamra, Abd el Melik, and Talh by the Tebus—were confirmed by natives of Kufra to be the three wadis that Wilkinson had previously heard about, and possibly they were the inspiration for the Zerzura legend. Penderel, a member of Almasy's expedition wrote, "Certain information we received from a Tebu guide in Kufara [*sic*] convinced us that there was still a third undiscovered wadi in the Gilf and that these three wadis must be the three wadis written of by Wilkinson in 1838." It is interesting that the third wadi appeared to have already been discovered by the native Tebu.[48]

Tebu knowledge of these wadis is discounted, relied upon, and then overcoded by colonial knowledge. Ironically, it is precisely when Zerzura is approached that it is discounted by the colonizers, becoming demythologized and thus internalized. Zerzura becomes a site of representational undecidability, or what Homi Bhabha has called a "space of translation," in which mapping and erasure are confused.[49] Almásy is in the desert specifically for the purpose of mapping its boundaries, but these boundaries are repeatedly conflated with the body. In *The English Patient,* Almásy says: "In 1930 we had begun mapping the great part of the Gilf Kebir Plateau, looking for the lost oasis that was called Zerzura. . . . We were desert Europeans. . . . And the Gilf Kebir—that large plateau resting in the Libyan Desert . . . —was our heart" (135). The reason Zerzura could not be found, then, is that when it gets too close, it becomes the body; when it is seen by the camera, it disappears, precisely because it had always been only a part of the legendary body of Herodotus.

Zerzura was never found but instead faded as an object of obsession, closing a period of frenzied and militaristic mapping of the Libyan Desert. This last blank spot of the earth began to be filled, as explorers began to both fear and lament the loss of their profession. Bagnold foresaw this future with sadness: "The time has come at last when the experts can close their notebooks, for there is nothing else unfound. We see Zerzura crumbling rapidly into dust. Little birds rise from within and fly away. A cloud moving across the sun makes the world a dull and colorless place."[50] Jean Baudrillard has said that postmodernism is "characteristic of a universe where there are no more definitions possible. . . . It has all been done. The extreme limit of these possibilities has been reached. It has destroyed itself. It has deconstructed its entire universe."[51] The closing of Zerzura could be said, then, to be the birth of postmodernism, when the world became a "dull and colorless place." This is the moment in which the colonizer is left with only himself, for there is nothing left to pursue in the world. The true representational crisis emerges, then, when the explorer finds he has destroyed his own heart; all the birds have flown away and there is nothing left but desert.

THE DISAPPEARANCE OF DOROTHY CLAYTON

The English Patient begins with an intentional misquote from the minutes of the Geographical Society meeting of "November 194–." It is an epigraph, which reads: "Most of you, I am sure, remember the tragic circumstances of the death of Geoffrey Clifton at Gilf Kebir, followed later by the *disappearance* of his wife, Katharine Clifton. . . . I cannot begin this meeting tonight without referring very sympathetically to those tragic occurrences." This quote is actually from a January 8, 1934, meeting in which the president of the Royal Geographical Society, Percy Cox, opened with remarks on a recent expedition to the Libyan Desert: "Most of you, I am sure, remember the tragic circumstances of the death of Sir Robert Clayton-East-Clayton, followed later by the *death* of Lady Clayton-East-Clayton. . . . I cannot call upon the readers of the paper to begin their address without referring, very sympathetically, to those tragic occurrences" (italics mine).[52]

There are several noticeable elisions in Ondaatje's rewriting of President Cox's remarks. The Clayton-East-Claytons have become the Cliftons of Ondaatje's narrative, and the meeting has been set forward at least six years in time to the 1940s. Lady Clayton-East-Clayton, also, is not dead but has disappeared. In *The English Patient,* Katharine Clifton's body is never found due to Almásy's capture and subsequent plane crash, turning her death into a disappearance. Dorothy Clayton's missing body provides the absent center of Ondaatje's narrative, a purposeful disappearance that serves to create mystery as well as function as that unmapped, nameless space that is repeatedly invoked in Ondaatje's story.

In *The English Patient,* Katharine is married to the explorer Geoffrey Clifton, who works with Almásy in the Libyan Desert. Katharine, because she is newly wed, insists on joining her husband on his expedition in the desert, where she proceeds to have a passionate affair with Almásy. Katharine's relationship with Almásy in *The English Patient* is less romantic than sadomasochistic. Almásy describes it in terms of the damage it has done: "A list of wounds. The various colors of the bruise. . . . The plate she walked across the room with, flinging its contents aside, and broke across his head, the blood rising up into the straw hair. The fork that entered the back of his shoulder" (153). Though the reason for the violence of their relationship is never completely explained, it is suggested that it is Katharine's obsession with ownership that makes their life together impossible. Almásy claims that his conflict with Katharine is based upon the fact that he is prone to disappear into the desert. Like the army of Cambyses, which in the legend is buried in sand,

Almásy claims he was sucked below the surface of the Libyan Desert: "Women want everything of a lover. And too often I would sink below the surface. So armies disappear under sand" (238). When Almásy at one point claims he hates ownership, her reaction is violent: "Her fist swings towards him and hits hard into the bone just below his eye. She dresses and leaves" (152). Even after they end the relationship, she says, "I still hate that about you—disappearing into deserts" (173). Katherine stands in contrast to Almásy in the sense that she comes from what Almásy calls "a fully named world"—she represents lineage, security, and the safe, ordered world of the English garden.

Interestingly, just as Ondaatje changed Almásy from a car salesman to a romantic, half-Bedouin desert trekker, so he also changed Dorothy Clayton, experienced pilot and explorer, to an angry lover and English housewife. In the same way, Dorothy Clayton's own attempts at exploration were elided from history, ultimately invoking and solidifying a patriarchal narrative of the Libyan Desert. Robert Clayton-East-Clayton (also known as Clayton), who had first discovered the wadi from the air, died before he could return with Almásy to enter the wadi by foot. Dorothy Clayton, after her husband's death, vowed to carry on his work.[53] She quickly organized an aerial expedition to the area of the suspected Zerzura site. The *Times* reported, "She took with her her husband's plans and maps, flew her own aeroplane, and travelled unarmed."[54] The Survey of Egypt office suggested that Dorothy Clayton should join the party of Patrick Clayton (no relation). Patrick Clayton had been organizing a separate expedition from Almásy's, because the members of the previous expedition had broken into rival groups. Francis Rodd lamented this rivalry in a letter to the Survey of Egypt office: "Almasi and Penderel are apparently going off on an independent trip rather before Lady Clayton will arrive. It seems to me, offhand, a pity that there should be two presumably rival expeditions to the same area. I am to suggest, on behalf of Lady Clayton, that Penderel should join her if Patrick Clayton agrees, so as to combine forces."[55] This combining of forces, for unknown reasons, was never to occur. Almásy left on an expedition with Kadar, Casparius, Bermann, two Sudanese drivers, and a cook. Even more discouraging, however, for Dorothy Clayton was the fact that Patrick Clayton also refused to wait for her. In a telegram to Francis Rodd, Ahmed Pasha Hassanein at the Survey of Egypt office wrote: "Clayton now in desert surveying part including wadi. . . . Penderel willing to help Lady Clayton flying after their trip or if she happens to be there same time. . . . Please assure Lady Clayton my sympathy and help."[56] While Dorothy Clayton had attempted to reunite all parties for a joint expedition of the wadi, she ended up having to travel with another explorer new to the area, Commander Roundell. Later, after Patrick Clayton's discovery of the first wadi, Dorothy Clayton and Roundell did join up with his group, though competition with the Almásy party remained keen. Dorothy Clayton wrote: "[Patrick] Clayton was anxious to make the journey through the middle of the whole length of the Sand Sea, a feat which would appear barely possible, but we accomplished it with really very little difficulty. . . . We had covered a wide area of unexplored country." One area of this unexplored country was the second wadi of Zerzura.[57]

Dorothy Clayton, nicknamed "Peter" by her friends, died at the age of twenty-six on a routine flight over Brooklands Aerodrome in England. Interestingly, she jumped from her airplane while it was taking off at fifty miles an hour, though the reason for this "accident" was never explained.[58] Three inquests deemed it "death by misadventure," even though eyewitnesses saw her jumping from the plane. The *Daily Express* carried the headline "Lady Clayton's Leap to Death." It suggested that "the only possible conclusion was that the throttle lever broke, and that Lady Clayton, acting on a sudden impulse leapt out without realising the speed of the machine." Still, Major Cooper, the air ministry expert, claimed, "The engine was quite controllable even with the lever broken, and it was extraordinary that she did not knock off the switches."[59] The airplane had been seen swerving to the left and to the right before she jumped, suggesting a broken throttle. At the inquest, the coroner asked Max Finlay, a flying instructor at Brooklands, whether a broken throttle could have caused the plane to crash. The coroner specifically asked if this could have been her reason for jumping: "Lady Clayton being in the back would have no means of stopping the engine?" But Finlay replied: "Yes, she would. There was the switch. All she had to do was to put her hand over the side and switch off." The coroner continued, "Did she know that?" to which Finley asserted: "Oh yes. She had used the switches."[60] So the reasons that Dorothy Clayton chose to jump rather than turn off the engine—if, indeed, the throttle was broken—may never be known.

Dorothy Clayton died on September 15, 1933, almost one year, to the day, after her husband's death. On September 16, 1933, her obituary in the *Times* focused on her husband's expedition work, as well as her "many and varied interests," ending with the fact that she was a "talented sculptor" whose home was filled with examples of her work. Even though the couple had publicly announced that they would "devote themselves to exploring" after their marriage, Dorothy Clayton continued to be described as a sculptor.[61] Though she accompanied her husband on his expeditions, the *Daily Express* reported after his death: "Sir Robert was married on February 29 to Miss Dorothy Mary Durrant, the twenty-five year old daughter of the Rev. Arthur Durrant vicar of Leverstock Green, Hertfordshire. She is a talented sculptress."[62] After Dorothy Clayton's death, the *Daily Express* remembered her only for her looks, reporting that she was a "young attractive widow . . . [who had] launched only a few weeks ago, [a] new hairdressing fashion: over and through her blond hair were a sleek natural looking lock of black."[63] The rare glass that Dorothy Clayton had discovered in the Libyan Desert and brought back to the Royal Geographical Society was similarly forgotten or displaced. The Natural History Museum, who had hoped to acquire these glass pieces (as well as Dorothy Clayton's photos) for exhibit, was disappointed to hear from Francis Rodd of the Royal Geographical Society that they had all been "lost."[64]

Dorothy Clayton's dismissal as a female explorer may be explained, in part, by the patronizing attitudes of her male counterparts. In a letter to Francis Rodd, a concerned fellow explorer wrote: "Thank you very much for letting me know that Lady Clayton has reached Egypt safely. I hope she won't be tempted to lose herself in the desert."[65] The concern expressed in this statement is that she might disappear—but not literally. Rather, the writer suggests that she might be seduced by the desert into a certain loss of control or even identity. Interestingly, this propinquity to be lost is also tied up in the Zerzura mystery, as Bagnold describes the circumstances surrounding the legend: "Always . . . the finder was at the time of discovery either himself lost or was wandering in search of a lost camel."[66] And yet, a woman's desire to "lose herself" is viewed as dangerous or even threatening. Women's supposed domesticity or vulnerability, it seemed, was not allowed for within the militaristic nature of the desert expeditions. For while these male explorers repeatedly expressed

a desire to unveil the romantic mystery of Zerzura, ironically, the flip side
to their pursuit was always militaristic. Lady Clayton-East-Clayton, who
notably traveled unarmed, stood in stark contrast to machine guns
mounted on cars.

ONDAATJE AND FASCISM

Since the release of director Anthony Minghella's film *The English
Patient,* both Ondaatje and Minghella have been accused of glamorizing
a fascist sympathizer. In 1940, World War II broke out in Africa. Almásy
joined the Germans and Italians in Libya, driving spies across the desert
from Libya into Egypt, just as his former colleagues were hired to cross
the desert, with machine guns, from Egypt to Libya. Bagnold formed the
Long Range Desert Group (LRDG), with the objective of carrying out
surveillance and raids in Libya; both William Kennedy Shaw, who had
previously worked with Bagnold as an explorer, and Patrick Clayton
were involved in these campaigns. Almásy worked for the Nazi general
E.J.E. Rommel and allegedly had a homosexual affair with him, as
reported by a nephew of Rommel who lives in Italy today.[67] In his obituary, Almásy was described as "a Nazi but a sports-man."[68] Arguments
both defending and accusing Almásy—and so Minghella and
Ondaatje—have centered on the question of Almásy's fascist associations. According to one critic: "Most sources about Almásy's activities in
the latter part of the 1930s and then during the war when he was with
the German army appear to agree that it is virtually impossible to establish whether Almásy was or was not a Nazi sympathiser although there
is evidence that he approved of Hitler's economic and social policies."[69]
After World War II, Almásy was tried by the Hungarian People's Court
as a war criminal but eventually released due to lack of evidence. He was
accused of collaborating with Rommel and writing a German propaganda book during the war.

In his defense, historian Peter Semeika argues, quite questionably: "If
he had any affiliations with the Nazis it was never more than a casual
matter which may have had its origins when his brother Janos and Adolf
Hitler were being simultaneously courted by the flighty British aristo-
crat, Unity Mitford."[70] It is not unlikely that Almásy similarly enjoyed
the company of such pan-fascist associations, though some critics claim
he moved in these circles only for mercenary reasons. Richard Bond

wrote: "He seems to me politically to have been a loyal Hungarian con-
servative serving an accommodationist government. As a scientific
explorer before the war he had worked for whoever paid best. Laszlo
Almasy could be described both literally and figuratively as a used car
dealer."[71] Describing the 1926 film of the African expedition of Almásy
and Liechtenstein, Raoul Schrott and Michael Farin write that, while
the film and the people in it exude the arrogance of colonialism, only
Almásy appears camera shy and detached.[72] This documentary film,
Across Africa in an Automobile, depicts Almásy as the quintessential engineer, building bridges, navigating with a compass, setting animal traps,
and endlessly measuring trophies. He is friendly with the natives, who
serve as guides and share hospitality with Almásy's group. Almásy, in turn,
shows them how to use guns and cameras and cars, stating, "Never before
have such wonder vehicles reached the inhabitants of the swamp islands."
At one point in the film, Almásy is measuring the antler of the rare
"swamp antelope." The caption of this silent-film image reads, "The precious prey in the camp," but when the action resumes, Almásy is this time
measuring the head of an African boy sitting on the antelope's back. This
simple displacement—boy as prey, boy as object of study—shows the
link in the European's mind between African and beast, as well as possible connotations to the boy as captured prey. One reviewer wrote,
"Though the documentary film is silent, it gives the impression that
Almásy loved killing animals and also that the explorers did not think the
locals of any more value than beasts."[73]

Ironically, Almásy was purportedly called "Father of the Sands" by
the local Bedouin. However, like the "Father of History," Herodotus,
who described the locals as deformed troglodytes, Almásy has an ambivalent fascination with the Bedouin. In *Across Africa in an Automobile,* the
desert is described as a land of "waterless death," creating an aura of
heroic risk around the activities of the gentleman-explorers. The desert,
it seems, must be empty of human life and sufficiently dangerous for the
men to demonstrate their courage and sense of adventure. The film does
not mention that about ten thousand Tuaregs lived in the southwest
desert, while about two thousand Tebu lived in southernmost Libya. In
Unknown Sahara, Almásy describes his attraction to the desert. "Some ask
me what profit humanity has from the exploration of the desolate sea of
rocks and sands," he writes. "Why waste money and risk your life for this?

I can only reply with an expression from the Bedouins: 'The desert is horrifying and merciless, but anyone must return to the desert who could ever comprehend it.'"[74] In this statement, Almásy is using the Bedouin to sanction his own vision of the horrifying desert; yet for many locals, the desert is simply home, not a merciless place to which one compulsively returns. In contrast to Almásy's bleak appropriation of Bedouin remarks, a Bedouin of the Sahara described the desert: "I will tell you something about the Sahara. This desert is very simple to survive in. You must only admit there is something on Earth larger than you. . . . You accept that, and everything is fine. The desert will provide."[75]

During World War II, Britain actively promoted the Senussi rebellion against Italy, guaranteeing independence for Libya if they succeeded. Mohammed Idris al-Mahdi al-Sanussi and Colonel C. O. Bromilow wrote to the Libyan leaders: "This is to inform you that the British government has decided to begin at once to organize battalions of the Sanusi Arab tribes in order to restore to them their liberty and emancipate their country from the hands of the Italian oppressors, and to insure their [country's] independence." In 1942, Anthony Eden made a similar statement: "His Majesty's government is determined that at the end of the war the Sanussi in Cyrenaica will in no circumstances again fall under Italian domination."[76] Lawrence would be disappointed by Britain's failure to keep such a promise in other Middle Eastern countries, but the promises to Libya were, in part, kept. Libya was not returned to Italy, though it became a joint British-U.S. mandate until 1951.

Ironically, while Libyans fought the Italians for the British, Rommel was applauded in Egypt as a liberator from British rule. Penelope Lively, in *Moon Tiger,* writes that Egyptians would put signs in their show windows that read, "German officers welcome here." World War II was a matter of indifference for the Egyptians, according to Lively: "When Egyptians speak of the war they mean the Israeli war, not ours, which hasn't after all anything to do with them."[77] The Senussi fought colonial authorities on both sides of the border, confusing the question of who the real enemy was. In *The English Patient,* confusion between the enemy and the self is also a common theme, as Caravaggio, Clifton, Almásy, and Kip all struggle with their national loyalties or conceal their true identities. Kip, an Indian working for a British bomb disposal unit, struggles between his loyalties to England and India. Clifton plays the role of an explorer and aerial photographer, but he is also an intelligence officer who secretly works for the British in the Libyan Desert. Caravaggio, who says he is a thief, also works as a spy for the Allies, though purely for mercenary reasons. He explains his usefulness: "We could read through the camouflage of deceit more naturally than official intelligence. We created double bluffs" (253). Almásy, when the Bedouin tribe first discovers him, muses: "During this time with these people, he could not remember where he was from. He could have been, for all he knew, the enemy he had been fighting from the air" (6). Counterpoised with this loss of identity, however, is the perpetual gaze of war, which leaves no detail unnoticed.

The camera became the ultimate arbiter of geographical placement or loyalty, and the ability to lose oneself in the desert became more and more impossible. The goal of the British was to quickly survey areas that might be of potential strategic importance during the war. Aerial photographs became a fast way of mapping out or verifying existing frontiers, and suddenly territories could be opened up that were behind enemy lines (see fig. 20). While the Almásy-Penderel expedition used the camera to photograph areas inaccessible by car, it was quickly discovered that aerial photography could prove essential for documenting areas inaccessible because of hostile forces in the area. Winchester and Wills noted that "the introduction of the aerial photograph completely changed the tactics of war. So much information that would otherwise have been concealed from the enemy was revealed by the all-seeing lens. . . . The camera automatically came to be recognized as the 'eye' of the Forces."[78] Before the invasion of Normandy, around two hundred million photographs were taken of the coast to help formulate the exact plan of attack. The feeling of being exposed in situations previously considered private permeated much of the war strategists' movements, as the eye of the camera replaced the eye of the explorer.

In *The English Patient,* the war imposes national boundaries on the desert, forcing the characters back into their respective national identities. In a sense, the novel is about decoding national identity, particularly that of the English patient, who turns out not to be English at all. Similarly, the photographing of Zerzura led to the elimination of that mythological space where identity could be lost. *The English Patient* romanticizes a prewar landscape that is also imperialistically European.

Ondaatje has been charged with aestheticizing fascism, but instead he fetishizes colonialism, demonstrating the way in which a postcolonial author may occupy a colonial space. This is physically evident in his occupation of the Royal Geographical Society, where he would have passed the life-size statue of David Livingstone to get to the journals of Almásy and Clayton.

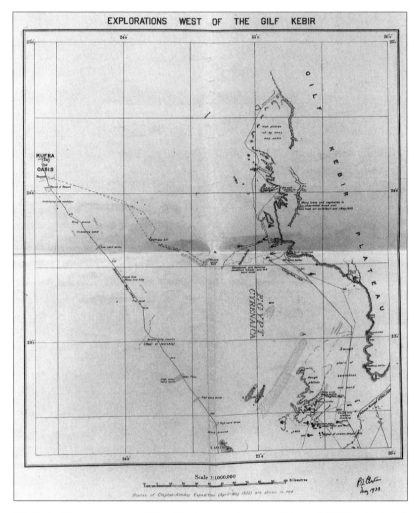

20. P. A. Clayton's map of the Clayton-Almásy expedition of 1932, which left from Kufra. Zerzura was thought to be located where it is noted that "many trees and vegetation" were seen from the air, mid- to upper right. Photograph from the Royal Geographical Society, London.

Ondaatje has been included in the canon of so-called postcolonial authors; he describes his family as "part Sinhalese part Tamil part Dutch part ass." He was born in Sri Lanka of Tamil and Dutch parents, though he emigrated to Canada at an early age. Ondaatje's memoir *Running in the Family,* which is set in postcolonial Sri Lanka, opens at an eighteenth-century Dutch governor's mansion in Jaffna, where Ondaatje's uncle Ned and aunt Phyllis are living. Ondaatje compares Sri Lanka's colonial years to a childhood fantasy: "From the twenties until the war nobody really had to grow up. They remained wild and unspoiled. It was only during the second half of my parents' generation that they suddenly turned to the real world." Ondaatje's family came from "two of the best known and wealthiest families in Ceylon," and his family's investment in colonial wealth and history is apparent throughout *Running in the Family.*[79] The Sri Lankan journalist Prakrti criticized *Running in the Family*: "This treatment of the political as once, twice or even thrice removed from reality, I read as an inability on the part of the writers who are located in an amorphous and transient cultural space . . . to grasp . . . all ramifications of the issues that plague the present Sri Lankan state."[80] Ondaatje's book *Anil's Ghost* addressed this type of criticism by focusing on the Tamil insurgency in Sri Lanka; however, this book is again mediated through the perspective of a Canadian woman, a forensic specialist hired by the United Nations to find evidence of human rights violations. The only Tamil in the novel is dead, and though his dead body is the focus of the book—as evidence—the book as a whole provides ample fodder for Gayatri Spivak's proclamation that the "subaltern subject cannot speak." By not representing the Tamils, the Tebus, or the Tuaregs, Ondaatje suggests that only transnationals can rise above the petty allegiances that lead to violence and war.

The heroes that Ondaatje continually chooses instead are wealthy hybrid (colonial/postcolonial) subjects. Aijaz Ahmad has noted this classist value system—which nonetheless denies its own class—in postcolonial literature: "All such *systems* [class, gender, nation] are rejected, in the characteristic postmodernist way, so that resistance can always only be personal, micro, and shared only by a small, determinate number of individuals who happen, perchance, to come together, outside the so-called 'grand narratives' of class, gender, nation."[81] It is precisely in this outside that Ondaatje repeatedly locates his characters, who are only truly free when they can remain in this limbo, together, fending off the encroaching "grand narratives" that threaten to overtake them and tear them apart. "A love story," Almásy claimed in *The English Patient,* "is a consuming of oneself and the past" (97). The desert provides the space for this consumption, where identity and nationhood can be erased, allowing love to occur. Almásy looks for Katharine out of a need to be consumed or loved, just as he hopes that Zerzura will consume or erase him. Similarly, Hana says of Kip, "She knows this man beside her is one of the charmed, who has grown up an outsider and so can switch allegiances, can replace loss" (271). To be one of the charmed is to be conjured, like Zerzura, out of the imagination, to be in that space where one is "lost from themselves . . . one more enigma, with no identification, unrecognizable" (95). And this is what it means to be a transnational subject, to take on—as Homi Bhabha has suggested— "hybridity as camouflage."

The Libyan Desert represents a site of hybridity in colonial discourse that, as it is opened up, becomes more and more representationally volatile or unstable. This is the process of translation, from Libyan to English, from colonized to colonizer, that Homi Bhabha suggests creates the hybridity and instability of colonial discourse: "The process of translation is the opening up of another contentious political and cultural site at the heart of colonial representation. Here the word of divine authority is deeply flawed by the assertion of the indigenous sign, and in the very practice of domination the language of the master becomes hybrid—neither the one thing nor the other."[82] This contentious political site is the center of the blank spaces on the maps, the empty center of the Gilf Kebir that Almásy claims "was our heart." If it is meant to represent the outside of European representation, it nonetheless is the home of the Senussi, who then must be symbolically erased; the problem, in colonial discourse, always is that someone is living there. Though Almásy calls the Libyan Desert the "land of waterless death," the Bedouin call it Tamazgha, the nation of the Imazighen people (commonly known as Berbers), which includes the cultures of Tuareg and Tebu. In 1995, the World Congress of the Imazighen met and declared their needs: "For the first time after hundreds of years of domination, assimilation, and resignation, Imazighen, from every corner of the world, came together to talk about one of their most pressing problems: the stake of their culture and language." The congress concluded, "Imazighen must write their own history."[83] Because the Imazighen have not had the opportunity to write their own history, their land has been depicted as outside representation—and so it is not the history of the Imazighen that Michael Ondaatje records, but the history of their colonizers.

Aijaz Ahmad claims that postcolonial authors today maintain a "growing ambivalence about nation and nationalism—combined with an even more surprising shift from a wholesale rejection of 'the West' to an equally wholesale assertion that the only authentic work that can be done in our time presumes (a) Third World origin, but combined with (b) metropolitan location."[84] Ironically, of course, most metropolitan locations are also cosmopolitan, or thoroughly Westernized. The struggle between the metropolitan and the rural, therefore, often serves to confuse the struggle between black and white. The metropolitan postcolonial author may face acts of racism, while at the same time adopting prejudicial attitudes toward rural or indigenous populations. It is precisely from this in-between space that *The English Patient* appears to be written. In an interview, Ondaatje complained that he was at first turned away from the Royal Geographical Society. "They weren't very friendly," he explained, though he was later let in. The *English Patient* film crew, on the other hand, was welcomed in immediately, which, he claimed, made the film more realistic than the book: "While most of it seemed dead on, it was more dead on than the book. It really bugged me when I heard . . . that they [part of the film crew] had gone to the Royal Geographical Society and were welcomed with open arms. I said, 'Weren't they difficult?' And they said, 'No, they were so nice.' "[85] Similarly, in *Running in the Family,* Ondaatje describes the English in Sri Lanka as "transients, snobs and racists" who "were quite separate from those who had intermarried and who lived here permanently."[86] In this sense, Ondaatje was made painfully aware that skin color could impact his treatment, and that of his parents. When the atomic bomb is dropped on Japan at the end of *The English Patient,* Caravaggio comments, "They never would have dropped such a bomb on a white nation" (286). Kip's Indian brother claims that he "sided with whomever was against the English" (291). Similarly, Ondaatje himself could be said to side with the Germans in this novel, or at least render their differences from the English more ambiguous than they might seem in the West. His uncritical depiction of fascism may be read as no different from the Egyptians' "German officers welcome here."

After the filming of *The English Patient* in Tunisia, Ondaatje described the days spent road building for the film: "And to have five hundred guys building a road in a desert! All these people!" One of the roads was named the "Saul Zaentz Imperial Highway," after the film's producer, and Ondaatje added: "Everybody who was there wanted to have a road named after themselves. Anthony wanted the Minghella Road, I wanted the Ondaatje Road. (laughs) The road that leads nowhere."[87] This statement may be said to reveal Ondaatje's feeling of exile or homelessness, which many critics claim is common to the post-colonial condition. But this metropolitan position tends to disavow the people living in the desert, building the road that leads nowhere. It also represents the colonization of the African landscape by the U.S. film industry, which may be no different from the English erecting statues of themselves in Africa, or Almásy and Clayton naming sand dunes and oases after themselves in the Libyan Desert.

The camera developed the third world with a speed thought impossible; yet that which was developed, from the air, was also often destroyed. This, in effect, is the basis of air control. In recent years, Major Marc Dippold suggested that air control had again become a viable strategy for control of the third world: "Reemergence of the issue of air occupation or air control is not surprising. The US economic 'empire' spans the globe—a world torn by increasing ethnic, religious, and nationalistic tensions. The task and costs of protecting our interests in this volatile environment are enormous."[88] The United States would again practice these air control techniques in Iraq during the Gulf War. Air control is the rationalized language of finding target tribes, writing expectations on leaflets, and following these messages with bombs. It is also the language of exploration—of Almásy, Lawrence, and Balbo—in which the strategy of recording/eliminating the colonized is practiced.

Michael Ondaatje does not mimic the rationalized rhetoric of these explorers; but by describing the desert and its peoples as a space where "identity and nationhood can be erased," he creates a space where colonization can occur. Ondaatje's explorers are caught up in a pan-European Africanist discourse, whose very intertexuality and internationalism is celebrated as an escape from a protofascist nationalism. It is ironic, then, that Ondaatje chose a fascist as his protagonist. In attempting to evade the pitfalls of nationalism, Ondaatje fell into the empty space of the desert; but in that space emerged fascism, which was, after all, a pan-European phenomenon. *The English Patient* is not for, or about, Libyans and Libyan struggles with the West. *The English Patient* is for Western eyes, through which, and indeed through rose-colored glasses, we may gaze at a romantic vision of ourselves.

PART THREE

Geographic Information Systems

CHAPTER 5

Canadian Cartography, Postnationalism, and the Grey Owl Syndrome

ON OCTOBER 24, 1946, the first pictures of earth from space were taken from a German V-2 missile fired from White Sands, New Mexico. The rocket ejected a movie camera, which floated to earth by parachute with the film running.[1] These V-2 rockets were acquired as part of Project Paperclip, a covert action in which 124 German rocket scientists and engineers—who had previously worked for Hitler or the SS—were brought to the United States to work for the Department of Defense.[2] In the 1960s, the advent of manned space flight meant a shift from covert action to publicity stunt, as "cosmonauts and astronauts in space capsules acted much like any tourists by taking photos out the window."[3] Soon the equivalent of TV cameras would be mounted on orbiting satellites, supplying low-level black-and-white imagery of earth. But because these satellites were U.S. controlled, the issue of who owned the photographs they acquired—the United States or the country being photographed—quickly became an argument over sovereignty.

In 1969, NASA began photographing Canada from its Earth Resources Technology Satellite. Alan Gottlieb, Canada's deputy minister of communications, objected to U.S. surveillance "on the grounds that the U.S. would be able to obtain exclusive information on the location of potential mineral and petroleum deposits in Canada by means of this satellite and might give advance information to U.S. exploration companies."[4] Canada accused the United States of invading its sovereign rights, primarily because territorial sovereignty had always been determined—in international law—by the ability of countries to "fix limits to their own sovereignty, even in regions such as the interior of scarcely explored continents where such sovereignty is scarcely manifested." The United States began mapping these areas with "scarcely manifested" sovereignty, and even today Canada is involved in a struggle over what this means for sovereign rights.

Canadian author Margaret Atwood, when asked what the ideal Canada would look like, responded: "First you cut along the border with scissors and you let the United States float away down near South America. That's the ideal Canada."[5] This attitude toward the United States stems from a prevalent anti-American vein in Canada, particularly after the implementation of the North American Free Trade Agreement (NAFTA). The United States has been accused of cultural and economic imperialism, as George Ball, the undersecretary of the U.S. Treasury, described the Canadian-American struggle in 1968:

> Canada, I have long believed, is fighting a rearguard action against the inevitable. Living next to our nation, with a population ten times as large as theirs and a gross national product fourteen times as great, the Canadians recognize their need for United States capital; but at the same time they are determined to maintain their economic and political independence. The position is understandable, and the desire to maintain their national integrity is a worthy objective. But the Canadians pay heavily for it and, over the years, I do not believe they will succeed in reconciling the intrinsic contradiction of their position.[6]

Sooner or later, Ball concluded, there would be free trade across the border; the implication, then, was that national sovereignty would crumble. Even before NAFTA, Canada had the highest level of foreign ownership (mainly by the United States) out of all industrial countries in the world. Canada is now required to grant unexamined takeover of Canadian enterprises worth between $5 million and $150 million. In an interview with Victor-Lévy Beaulieu, Atwood voiced her dissent: "There's a clause in the free-trade agreement that stipulates that, if the government of Canada grants money for the arts or education exclusively to Canadians, the United States can take reprisals in any industry."[7] This, according to Atwood, represents the beginnings of cultural imperialism.

Margaret Atwood has always been critical of U.S.-Canadian relations; she suggested in 1972 that "Canada as a whole is a victim." This attitude, she claims, is largely due to Canada's awareness of itself as a colony, "a place in which profit is made, but *not by the people who live there:* the major profit from a colony is made in the centre of the empire."[8] Interestingly, Atwood never explicitly defines the center of empire as England. Instead, she suggests that it is the United States that profits, revealing the complex position of Canada in any discussion of postcolonialism. Bart Moore-Gilbert, in *Postcolonial Theory,* suggests that there are five possible contexts in which this term might be applied to Canada: its original dependent relation to Britain, its current dependent relation to the United States, the secessionist movement in Quebec, the colonization of the First Nations, and the substantial racial minorities that have immigrated since the end of the Second World War.[9] The Quebeçois (French-speaking Canadians) colonize the First Nations as the English-speaking Canadians colonize the Quebeçois as the Americans colonize the Canadians.

In *Survival,* Atwood attempted to define Canadian literature in contrast to its American neighbors and British ancestors. In her definition, Canadian literature contains the motif of survival more than its Anglo cousins do because of its complex history of dependency on both of them. In an interview, she explained:

> I ended *Survival* with the question 'Have we survived?' That question is still very current. I wrote the book almost twenty-five years ago, and since then our economic domination by the United States is even greater. The Canadian government is dismantling all its structures for supporting the arts, social programs, and health care system, care of the elderly, and so on. The question of Quebec's separation also has a potential impact on this. Canada will be transformed into several little countries.[10]

What Canada is facing today is the end of nationalism in a much more literal sense than postnational theorists of global capitalism would suggest. Besides the economic domination from the United States, Canada has the potential to actually splinter into different countries. If Quebec secedes, the First Nations of Quebec claim that they will either stay with Canada or seek their own independent state. And if they secede, what of other territories with ethnic majorities, such as the recently created First Nations territory of Nunavut?

Postnationalism is generally considered a kind of transnational cooperation between previously opposed peoples. Leela Gandhi described it as "a new transformation of social consciousness which exceeds the reified identities and rigid boundaries invoked by national consciousness."[11] NAFTA, in essence, represents the emergence of postnationalism. It also, however, represents the end of cultural sovereignty. The economic and cultural colonization of Canada by the United States is merely the next step in a long history of colonization: First Nations, French, English, Canadian, American. What "sovereignty" means, for anyone, in the face of this, is of urgent concern, and it is a question that the rhetoric of postnationalism appears incapable of answering.

THE HISTORY OF GEOGRAPHY

"Canada, they say, has more geography than history," Roger Tomlinson said in an interview for *GIS World*. Tomlinson is generally considered the father of Geographic Information Systems (GIS), which was invented and first used in Canada in the 1960s.[12] This common saying about Canadian geography reveals the country's colonial roots. Canada, like the United States, does not have history—that is, in a European sense—because the settlers have not been there long enough. But unlike the United States, 80 percent of the Canadian population lives one hundred miles or less from the border, making mapping of the sparsely inhabited North the only way to establish sovereignty and administer its vast territory. To provide this function, Canada has always invested heavily in cartographic operations—particularly aerial photography—and today Canada has one of the largest aerial photographic libraries in the world.[13] The very immensity of Canada's photographic library led to the need to compile these resources in a computer database for analysis and preservation purposes. By the early 1960s, the Canada Land Inventory project (CLI) was already using GIS to develop "an inventory of all the land across the country that was under farming practice or could be turned into forest plantations or some alternative use."[14] The purpose of CLI was to enable development of remote regions, labeling them for four uses: agriculture, forestry, recreation, and wildlife. Land was ranked on a scale of one to seven for each of these potential uses, "independent of location, accessibility, ownership, distance from cities and roads, and present use of the lands." But, CLI suggested, "it does assume good management."[15] What is left off the maps allows CLI to implement its vision of "good management." CLI admits that "the information was generalized during air photo interpretation to fit the classification." The maps were compiled on a scale of 1:50,000 (1 centimeter = 0.5 km), so that small-scale or microclimactic developments were not included in the CLI-GIS vision of the world.

CLI contended that one of its objectives was to create "a historical document presenting the use of the land at a particular point in time."[16] But the only categories that were included on its land-use maps were recreation, mines, cropland, wetlands, productive woodland, nonproductive woodland, improved pasture, rough grazing, intensive cultivation, and "built-up" (i.e., cities). Lands that did not fall into these categories were labeled "barren." Lands where the Algonquin live were called productive woodland, Kipawa lands were considered recreation areas, and—farther north—the rocky Innu territories were designated barren.

The land of the Kipawa and Algonquin is where Margaret Atwood was raised. She writes of her fear of surveillance in *Surfacing,* where the narrator describes watching a boat with an American flag on her lake: "I thought they were going to land: but they were only gazing, surveying, planning the attack and the takeover."[17] There are also mysterious surveyor flags around the lake, which she expects is for "raising the lake level" again. Atwood fears interference in this area of northern Quebec from the United States or the Canadian South, but she also claims to have a feeling of being "home" only through an aerial perspective of the North:

> I'm most at home in an airplane, a thousand feet up, skimming over the taiga at one remove. Lake, lake, lake, swamp, sprinkle of low hills, twist of river; ice creeping out from the shores. It has to be big, though; rocky, sparse, a place you could lose yourself in easy as pie, and walk around in circles and die of exposure. . . Out the window, way down there: desolation, instant panic. With a view like that you can feel comfortable.[18]

Atwood invokes the feeling that, in a settler society, one hovers above, wanting to be a part of the landscape, but knowing at the same time that it causes only "desolation" and "panic." Atwood appears to see, like CLI, a barren and unproductive landscape in the North. In fact, all it produces is death, which is part of its appeal.

In a series of lectures that she gave at Oxford, Atwood chose the Canadian North as her topic. She explained her choice: "Canadian literature as a whole tends to be, to the English literary mind, what Canadian geography itself used to be: an unexplored and uninteresting wasteland, punctuated by a few rocks, bogs, and stumps." What Canada did have to offer, and what defined its literary success, she ultimately determined, was the North: "Popular lore, and popular literature, established early that the North was uncanny, awe-inspiring in an almost religious way, hostile to white men, but alluring; that it would lead you on and do you in; that it would drive you crazy, and, finally, would claim you for its own."[19] Atwood explained that her ideas about the Canadian North also formed the symbolic nucleus for her own work, including *Surfacing, Wilderness Tips,* and her poem sequence *The Journals of Susanna Moodie.*

Surfacing, like Susanna Moodie's *Roughing It in the Bush,* has been called a woman's *Heart of Darkness.* Mary Lannon wrote that Atwood "tells a quest-story qua *Heart of Darkness* from a woman's point of view. Her jungle is not the African other's but her own home turf . . . the evil [is] always outside of her in the form of Americans or men."[20] *Surfacing* is the tale of a Canadian's journey into the wilderness in an (aborted) attempt to make it home. The narrator goes to a remote island in northern Quebec in search of her missing father, then gradually goes mad (she plans to grow fur) and decides to hide from her friends in order to remain alone on the island. When her friends return to search for her, she sees them as cyborgs, claiming that they "are already turning to metal, skins galvanizing, heads congealing to brass knobs, components and intricate wires ripening inside" (160). She believes that they have come for her because she would not sell her land and claims not to know them or recognize their language, saying they may be police, or tourists, or an "invasion" from America.

"America," in the novel, is representative of anything that is invasive, inhuman, or technological. It suggests a confusion over nationality that parallels the sovereignty issues emerging with the introduction of satellite imaging and multinational collaborative technologies. "It doesn't matter what country they're from," the narrator explains, "they're still Americans, they're what's in store for us, what we are turning into. . . . Like the late show sci-fi movies, creatures from outer space, body snatchers injecting themselves into you dispossessing your brain, their eyes blank eggshells behind the dark glasses" (130). Similarly, she describes her friends as being taken over with "second-hand American," which is the reason she runs from them. "Behind me they crash," she thinks, "electronic signals thrown back and forth between them, hooo, hooo, they talk in numbers, the voice of reason. They clank, heavy with weapons and iron plating" (191). She hides in the swamps from them, trying to protect herself and the child she believes she carries.

In the end, the narrator decides to be human and return to civilization. Critics tend to read her return as an acceptance of human "responsibility" and thus a victory.[21] Instead, I would argue that her return is based on a recognition that it is impossible to protect oneself from foreign invasions, whether from men or cyborgs or Americans. Though she goes to the wilderness to escape all of these things, she realizes that the wilderness itself has been invaded by the United States. Throughout the novel, there are suggestions of possible CIA involvement in the area, even suggesting that the narrator's father could have been killed by the CIA. "Perhaps the C.I.A. had done away with him to get the land," the narrator's friend says, believing that the U.S. government would stop at nothing to gain access to Canada's resources (101). The island is surrounded by underground U.S. military bunkers used during World War II. The narrator comments: "I heard they'd left, maybe that was a ruse, they could easily be living in there, the generals in concrete bunkers. . . . There's no way of checking because we aren't allowed in. . . . That's where the rockets are" (5). And when a U.S. citizen—with a foreign accent and a German name—comes to buy the island for the "Wildlife Protection Association of America," the narrator's friend instantly suggests, "He was a front man for the C.I.A." (96). The island functions as a Robinson Crusoe–like setting for a test of the narrator's character, but it is also—unlike Crusoe's island—being watched. She returns to the city, in the end, in order to find the people who are taking her land, making the maps that shape her territory—to not be a victim anymore.

In *Surfacing,* the narrator's father moves his family from southern Ontario to northern Quebec in order to escape the effects of war and the United States. Even in the wilderness, however, there are projects occurring on the lake about which the narrator can only conjecture. She imagines that "surveyors, the paper company or the government, the power company . . . were going to raise the lake level" (114). Her father, she realizes, is somehow a part of these development schemes, whose residue can be seen in the surveyor flags and "advance men, agents" doing fieldwork. She suggests, "Our father had gone on a long trip . . . for the paper company or the government, I was never certain which he worked for" (78). Later, she says that her father was "really a surveyor," mapping the terrain for forestry or dam projects. Americans, it seems, are somehow behind this conspiracy, though they can never truly be apprehended for "they whoosh away into nowhere like Martians in a late movie" (64). The island is thus continually stripped of local meaning; it is controlled from elsewhere by the hidden navy base, the mysterious paper company, and the reservoir surveyors.

The indecipherability of the Canadian landscape is a recurrent theme for a long line of settler women, from Susanna Moodie to Margaret Atwood. For Atwood: "We are all immigrants in this place even if we were born here: the country is too big for anyone to inhabit completely, and in the parts unknown to us we move in fear, exiles and invaders. . . . This country is something that must be chosen—it is so easy to leave—and if we do choose it we are still choosing a violent duality." For women, who were so often following their husbands into the wilderness, there is also the sense of being stranded, forced into exile in a foreign environment. There is, in Canadian women's settler literature, a sense of trying to find a compromise—make a home—with a wilderness that can never be home. This is in direct contrast to their male counterparts, who, as Atwood suggests, tend to be pragmatically focused on the outside as a place to make a living, or an opportunity. Women's narratives, instead, often evoke a sense of siege. In Margaret Laurence's *Stone Angel,* the physically deteriorating heroine, Hagar Shipley, escapes to a remote cabin to avoid relatives whom she cannot endure. If Hagar Shipley feels besieged or haunted by relatives—or her own infirmity—in the wilderness, Susanna Moodie feels perpetually threatened by wild animals and Indians. Carol Shields suggests that Susanna was "a Crusoe baffled by her own heated imagination, the dislocated immigrant who never fully accepts or rejects her adopted country." Atwood has famously said that the national illness of Canada is "paranoid schizophrenia," suggesting that Susanna Moodie embodied this model. In *The Journals of Susanna Moodie,* Atwood attempts—in a series of narrative poems—to fill in the emotional gaps of Moodie's own *Roughing It in the Bush.* In so doing, she turns

Moodie into a veritable Kurtz: "We left behind one by one the cities rotting of cholera, one by one our civilized distinctions and entered a large darkness. It was our own ignorance we entered. I have not come out yet."[22]

Settler women, however, are no less condescending to the natives than are their husbands. If anything, the fear seems to be greater, as the possibility of different cultural and sexual practices appears to forever haunt the dislocated imagination of the settler. Moodie claims that the character of the Indian is "dark and unlovely" and "invested with a poetical interest that they scarcely deserve." Of the Chippewa in her region, she writes that they are "perhaps the least attractive of all these wild peoples, both with regard to their physical and mental endowments." Ironically, it is the Chippewa that keep her family alive during the long winters, often bringing "a fine bunch of ducks" or putting a "quarter of venison" at the door when Moodie "had nothing to give them in return, when the pantry was empty, and 'the hearth-stone growing cold.'"[23] Atwood highlights Moodie's sense of fear in her *Journals,* writing: "The forest can still trick me: one afternoon while I was drawing birds, a malignant face flickered over my shoulder. . . . Resolve: to be both tentative and hard to startle (though clumsiness and fright are inevitable) in this area where my damaged knowing of the language means prediction is forever impossible."

There is a scene in Susanna Moodie's *Roughing It in the Bush* where the Chippewa find a "large map of Canada" in Moodie's house and are "infinitely delighted." According to Moodie: "In a moment they recognized every bay and headland in Ontario. . . . What strange, uncouth exclamations of surprise burst from their lips as they rapidly repeated the Indian names for every lake and river on this wonderful piece of paper!" Though the Chippewa want to purchase it, Moodie declines on the grounds that "the map had cost upward of six dollars, and was daily consulted by my husband."[24] This account contradicts the stereotype of indigenous people who are unable to read maps of their own territory or even understand its parameters. Barry Lopez similarly writes of the Inuit that, besides producing their own maps, "they could also read European maps and charts of their home range with ease, in whatever orientation the maps were handed to them—upside down or sideways. And they had no problem switching from one scale to another or in maintaining a consistent scale in a map they drew." Moodie's narrative shows this instantaneous understanding, though the Chippewas' desire for the map is considered not as important as her husband's need to learn the "names and situations of localities in the neighborhood."[25] Though the Chippewa instantly overcode the map with their own names, Moodie understands that the importance of the map is that her husband learn the European names and so be able to locate himself. Interestingly, Moodie never suggests that she herself can read or understand the map but simply recognizes its economic and colonial values. In this sense, she is doubly lost.

THE GREY OWL SYNDROME

"It was one of the accepted truisms of Canadian literary criticism in the 1950s and 1960s," Atwood writes, "that the Canadian poet's task was to come to terms with the ancient spirit—that is, the Native spirit—of the land whites had not yet claimed at a deep emotional level." The goal has been, she suggests, to claim the natives as ancestors, though she admits that "the fact is that they may not particularly *want* to be our ancestors." This conjectured resistance to the white claims for kinship, however, must ultimately be ignored. She says: "But like it or not, the wish to be Native, at least spiritually, will probably not go away; it is too firmly ingrained in the culture, and in so far as there is such a thing as Canadian cultural heritage, the long-standing white-into-Indian project is a part of it. . . . Perhaps we should not become less like Grey Owl and Black Wolf, but more like them."[26]

Grey Owl was a famous Canadian who emigrated from England in the 1920s and successfully went native, eventually being adopted by Ojibwe Indians. He then not only wrote articles and a highly successful book in his native persona, but also passed as native on a speaking tour in England, even convincing the king he was Indian. Atwood confessed to her own desires to go native during her appearance at Oxford: "I thought of pulling a Grey Owl impersonation, turning up in buckskin and a feather head-dress and beginning, 'I come in peace, brother.'" Part of Atwood's native ruse was a desire to play the primitive for an audience that, she admits, was tired of Shakespeare and "was very fond of cannibalism."[27] It is ultimately only the gaze of the British that made Atwood want to play out colonial stereotypes about indigenous people;

in so doing, she could claim the indigenous space for herself, making Canadian literature authentic and native. In this sense, it could be said that the British, not the First Nations, might have inspired *Surfacing*. At the same time, there is, in Atwood's work, the perpetual "feeling of being alien, of being shut out, and the overwhelming wish to be let in."[28]

In *Surfacing,* this desire comes across in the father's aborted attempt to go native or understand Indian spirituality; he does it, interestingly, by making a map. The father starts a search for indigenous pictographs, which he then begins to map, hoping this will help to atone for his past deeds as a forester-surveyor-scientist: "He has realized he was an intruder. . . . He wants it ended, the borders abolished, he wants the forest to flow back into the places his mind cleared: reparation" (192) He hopes to eliminate the corporate-state map, which has drawn the borders of his territory, turning instead to indigenous maps of the territory:

> The Indians did not own salvation but they had once known where it lived and their signs marked the sacred places, the places where you could learn the truth. . . . He had discovered new places, new oracles, they were things he was seeing the way I had seen, true vision; at the end, after the failure of logic. When it happened the first time he must have been terrified, it would be like stepping through a usual door and finding yourself in a different galaxy, purple trees and red moons and a green sun. (146)

The father becomes a kind of indigenous visionary in the novel. He begins on a project of mapping out native rock paintings, which become an elaborate code for his daughter to solve. The narrator finds a map in her father's cabin covered in Xs. She soon realizes that the numbers on the sketches correspond to numbers by the Xs on the map, and that they must designate rock paintings that he was sketching. She also discovers a note from an academic who has tried to identify the paintings, suggesting that they mark the abodes of a "powerful or protective spirit" and are linked to indigenous rituals (103). But when the main character goes to find these paintings, she learns that her father had started by sketching native paintings but later began to make his own paintings and experience visions of his own. By wedding his own markings with indigenous markings, he confused or bridged the gap he felt in an alien territory.

The father drowns while trying to photograph the pictographs, the weight of the camera supposedly dragging him under the water. His daughter, in turn, begins to follow in his footsteps, trying to decode his maps and so discover his whereabouts. Searching for him, she comments, "He had been there and long before him the original ones . . . the first explorers, leaving behind them their sign, word, but not its meaning" (127). Just as she cannot decipher the meaning of her father's pictures, so the indigenous pictures are said to be indecipherable—they are the terrifying secret at the center of the landscape. But by calling indigenous people the "first explorers," the narrator creates a direct lineage between them and her father. Like him, they were explorers, and like him, they charted cryptic maps of the territory. The supposedly indigenous paintings in the novel are also described in terms of the paintings that she and her brother drew as children—dismembered soldiers, man-eating plants, and "people-shaped rabbits"—rather than in terms of actual native pictographs. The brother is said to have drawn "purple jungles . . . the green sun with seven red moons," and the narrator claims that the father's vision must have been like entering a galaxy with "purple trees and red moons and a green sun" (90, 146). The desire to go native and have indigenous visions of the land becomes confused, in the narrator's mind, with the desire to reclaim her own history and childhood. She learns to read her own childhood drawings as pictographs, happy that her mother saved them for her. She comments: "They were my guides, she had saved them for me, pictographs, I had to read their new meaning with the help of the power. The gods, their likenesses: to see them in their true shape is fatal. While you are human; but after the transformation they could be reached. First I had to immerse myself in the other language" (159). Like her father's learned ability to mimic native pictographs, the narrator learns that she had drawn pictographs already—a fact that she can only see when she has been transformed, halfway between animal and human.

The narrator's quest for her father leads to her gradual rejection of her friends, her clothing, and ultimately her mind. She says: "I no longer have a name. I tried for all those years to be civilized but I'm not and I'm through pretending" (173). Later she sees herself in a mirror and describes herself as "a creature neither animal nor human" (196). In this transformation there is a bizarre conflation not only between native and non-native but also between animal and human. Going native, with all

its primitivistic associations, means regressing to a supposedly more natural state in which the Indians are forever arrested.

It is no wonder, then, that real Indians are generally of no concern
to Margaret Atwood. In *Surfacing,* there are no Indians, only whites who
go native. Similarly, Atwood dismisses the impact of the native Algonquin
in her home in northern Quebec outside Témiscamingue. When asked
about the Indians, she said that "a lot of them were killed by tuberculo-
sis at the end of the nineteenth century, and by influenza." Instead, she
claimed, the area was full of "lumberjacks and railway workers."[29] But
today the Algonquin have outstanding land claims in the area. The locals
of Témiscamingue, now faced with a dying forestry industry, have also
expressed their discontent with Atwood's understanding of them. Mayor
George Lefebvre, at a local "unity rally," claimed that nonlocals did not
understand their lives: "They have no grasp about what life in the North
is really about." Lefebvre claimed that his message also went out to Margaret Atwood—the message, he said, was "go home." Individuals at the
rally held up signs that read "GO HOME" in capital letters, which the
group also began to chant.[30]

The problem with Atwood, indeed, is that she never conceives of the
Canadian North as "home." Instead, it is a place of both spiritual renewal
and terror, a femme fatale in male explorer narratives that, when confronted by women, becomes a kind of sister. Atwood becomes invested
in reclaiming the "gone native" narrative for women in an effort to com-
plicate and explore a previously patriarchal genre. "There's . . . a shortage
of stories about women who wanted to turn themselves into Indians, like
Grey Owl," she writes. "Pretending you're part of a hunter-gatherer society is not much fun if all you get to do is the gathering." In her effort to
do this, however, she simply reaffirms primitivistic stereotypes of the
Indian, asking, "If there's any cannibalism to be done, what happens
when it's women doing the eating?" Yet there is also a certain sense of
parody to this project, a tongue-in-cheek witticism of dressing up as a
man who dresses up as Grey Owl. This camouflage could be said to actually undermine the glory that was recently reaffirmed in the film *Grey
Owl,* starring Pierce Brosnan. There is also the sense of performing the
native for the British, understanding their complex and absurd desires for
cannibalism. But ultimately, Atwood is seeking a performance that will
temporarily alter, disorient, or challenge the parameters of her character's

identity, even though, in the end, she always returns home, to the city and civilization. She writes, "A sort of feminist enterprise gets entwined with the older Seton-and-Grey Owl project of turning the wilderness into a spiritual health spa."[31] Atwood's "feminist–health spa" narrative betrays her own class interests; the wilderness is, after all, nothing more than an expensive retreat for those who do not live there. Grey Owl's outfit is ultimately just pleasant and eye-catching costume garb for an evening out at Oxford.

THE SOVEREIGN SOUTH

In *Surfacing,* there is the perpetual sense of control being located elsewhere: "My country, sold or drowned, a reservoir; the people were sold along with the land and the animals, a bargain, sale, *sold* . . . the flood would depend on who got elected, not here but somewhere else" (133). This is the reason that the narrator must return, claiming that she "must refuse to be a victim" (197). Remaining in the wilderness, she is still a victim of the power company and the surveyors and the American hunters. After her transformation, she feels she can return to the city and the "Americans" without becoming their victim again: "They exist, they're advancing, they must be dealt with, but possibly they can be watched and predicted and stopped without being copied" (195). Control is located in America, where she ultimately returns. In this sense, she does "GO HOME," though in so doing, she may alienate herself from those who remain.

Marian Engel's *Bear* also relates the story of a woman, Lou, who moves alone from the city to a remote northern island, only to return after a spiritual transformation. Lou engages in sexual encounters with the pet bear of the previous owners on this island, symbolically rejecting her troubled relation with men and the patriarchal order by allowing— even training—the bear to perform oral sex on her. Though her relationship with the natural world is eroticized in this fashion, her actual attempt at bestiality fails when the bear, remembering his own wildness, tears her back with his claws. The boundaries between nature are thus violently redrawn as Lou returns to the city with a mark—like *Surfacing*'s unborn child—that empowers her to face her enemies. These two novels demonstrate a fascinating experiment in recovering the local, the sexual, and the indigenous natural world for women; both attempts at localization fail, however. In *Surfacing*, the narrator is perpetually *anglais* in a French world, white in a native world, and human in an animal world. In *Bear*, Lou's lust is mingled with an extreme sense of depravity until the bear violently rejects her humanity and returns to its own world. In both novels, there is a pervasive sense of the impossibility of disappearing in this isolated outpost from civilization. "Americans" are encroaching, as are surveyors, tourist cottages, and wilderness societies.

Atwood and Engels empty the North of its native inhabitants, resurrecting them only as talismans or guides for their own spiritual discovery. In *Surfacing*, these talismans are the native pictographs. In *Bear*, Lou is at first unable to get close to the bear; inexplicably, a one-hundred-year-old Indian woman briefly appears to explain how to befriend the bear. "Shit with the bear," she says. "Morning, you shit, he shit. Bear lives by smell. He like you." By defecating where the bear can see her in the morning, Lou gradually wins the trust of the bear; soon, he is regularly providing her with orgasms. Ultimately, when she realizes her difference and decides to return home, she does so with a sense not of defeat but of victory. She has the marking of the bear and claims that "she was at least clean."[32] In the end, Lou returns only with a copy of John Richardson's *Wacousta*, a narrative of a man who disguised himself as an Indian.

In *Surfacing*, the narrator claims to gain her power or vision from the maps she draws and collects, and ultimately she returns to the city to apply this newfound power. Though she associates this power with a native vision, what she has actually acquired is a map. The narrator could not "be an Indian" in the way that Indians define themselves; her own unawareness of this fact only reveals her colonial tendencies toward a primitivist view of natives. Instead, she hovers above, like the Americans, unaware of what is producing her own panic, which is simply the space of an unknown Canadian wilderness and the fact that she would be lost within it.

It is this space that provides the impetus for mapping the "scarcely manifested" sovereignty of the North. The anxiety that emerges in the face of this space is, indeed, that it is not home. The Canadian North led to the development of GIS, in order to discover it, manage it, and map its resources. This includes establishing data-gathering procedures, systems of measurement, and map categories. In the Land Inventory projects, for instance, the maps only work because of the consensus reached over the kinds of labels that would be applied to the territory. The CLI project, like national atlases around the world, establishes a sense of ownership, shared values, and the ability to create a centralized plan. CLI is written in the language of nationalism.

In order for this type of national discourse to occur, however, something must be erased. The "perennial domination over nature" is made possible "only by the process of oblivion," Horkheimer and Adorno wrote in *Dialectic of Enlightenment*. "The loss of memory is a transcendental condition for science. All objectification is a forgetting."[33] This occurs, in cartography, through a process of becoming lost, erasing the past in order to make a map or an objective representation of the territory. In the wilderness, then, that liminal position between being lost—or losing oneself—and making the map represents the quintessential colonial desire. Frederick Jackson Turner defined this process most clearly:

> The wilderness masters the colonist. It finds him a European in dress, industries, tools, modes of travel, and thought. It takes him from the railroad car and puts him in the birch canoe. It strips off the garments of civilization and arrays him in the hunting shirt and the moccasin. . . . Before long he has gone to planting Indian corn and plowing with a sharp stick, he shouts the war cry and takes the scalp in orthodox Indian fashion. In short, at the frontier the environment is at first too strong for the man. He must accept the conditions which it furnishes, or perish, and so he fits himself into the Indian clearings and follows the Indian trails. Little by little he transforms the wilderness, but the outcome is not the old Europe. . . . The fact is, that here is a new product that is American.[34]

Though Turner was discussing the American West, he claimed that the same process occurred in the South, the Far West, and the Dominion of Canada. The European settler wants to throw off his history in order to claim this new territory as his own. In throwing off Europe, the tendency is to go Indian in order to claim indigenous space. But the very anxiety about that position, which is equated with madness or duplicity, compounded with the simple terror of perpetually being lost and potentially unsafe, establishes the difference of the settler.

Instead, a portable environment is required, a simulation of the North that can be brought back to the South. This is precisely what GIS supplies. Going native, in the colonial Canadian explorer sense, serves the double purpose of overcoding the actual natives with your own presence and erasing their territory for your maps. "GIS," Richard Jonasse wrote, "is a tool for the creation of factual, portable landscapes that are used to make sense of the world, calculate/model different scenarios, and control activities in remote locations. . . . GIS is predicated on the many generations of activity that went into making the world amenable to representation." GIS makes the world portable. Jean Baudrillard, in *Simulations,* described the contemporary problem of "the *satellization of the real,*" which he called "the putting into orbit of an indefinable reality." When what is real becomes only what is portable, then vast areas of situated knowledges are devalued and disappear, primarily indigenous ones.[35]

This mechanism is something of which First Nations people have long been aware. "Indigenous peoples have historically reviled attempts by central governments to collect information on them," Wayne Madsen explains. "Subjecting indigenous peoples to remote space-based imagery and surveillance without proper guarantees of their rights to privacy and self-determination can only exacerbate strained feelings."[36] GIS critics Peter J. Taylor and Ronald J. Johnston discuss the "social relations" expressed in satellite data sets, suggesting, "The basic relation is that a person (A) decides to collect information on another person or thing (B)." This constitutes a power relation in which "the more powerful do the finding out about the less powerful." For this reason, they claim, "A as owner of information about B can decide what to do about B." And yet, these collectors of information "do not find out everything, since A selected what it is about B that is useful to collect."[37] As control is given over to A, everything that is not considered useful about B is discarded or forgotten.

Multinational corporations have used this approach of discarding native information in order to target their lands for exploitation. The effect has been that First Nations groups have been forcibly evicted from their lands and forced into unfamiliar urban settings. In Quebec, the most flagrant example of this was inundation of the James Bay area by Hydro-Québec, including the lands and villages of nearly ten thousand Cree. Though the Cree tried to stop this project, the court ruled in 1973 that Charles II of England had granted rights to the territory to the Hudson's Bay Company in 1670. In 1975, the Cree demanded compensation for their loss, resulting in the James Bay and Northern Quebec Agreement. The agreement awarded $225 million to the Cree and guaranteed them the right to govern, hunt, and fish on their traditional lands. Ironically, this agreement is now being used against the Cree by the Quebeçois who want to secede—they claim that the Cree gave up their rights to their land by signing the agreement and therefore would not be allowed to remain with Canada or become independent. Matthew Coon Come, the grand chief of the Cree, told reporters: "Quebec independence is the greatest threat to the Cree since the arrival of Europeans in North America. . . . But if Quebecers have the right to self-determination, certainly the Cree do. We cannot simply be traded from one country to another as though we were cattle."[38] The Cree in 1995 held a referendum in which 96.3 percent voted to stay with Canada.

Matthew Coon Come also successfully campaigned to stop an extension of the James Hydro Project using GIS. Hydro-Québec had claimed that the lands to be inundated were unused; the Cree, in turn, successfully used GIS to represent their own residency. The maps that the Cree consolidated in GIS were clear evidence of continual use and occupation of their lands, all of it invisible to Hydro-Québec. In 1995, Wade Cachagee of the Chapleau Cree started Cree-Tech, a GIS company specifically designed for First Nations uses. Cree-Tech can produce a "traditional values inventory" that maps out traditional farming, trapping, fishing and hunting areas; pictograph and burial sites; traveling routes; and periodic and permanent settlements.[39] In its first year, Cree-Tech mapped the coast of James Bay. The group called Six Nations—including Mohawk, Oneida, Onondaga, Cayuga, Seneca, and Tuscarora—also developed GIS software, Eagle's Cry. President Philip Monture remarked: "The idea of connecting information to the land is not a new concept, and Six Nations does not profess to be presenting a new theory. What Six Nations Geo Systems has done is to take its traditional beliefs . . . and adapt them using new technology to benefit its people for today and our children for the next seven generations."[40]

First Nations, today, are one of the faster-growing groups of GIS users. More than half of Canada's land base is subject to land claims by First Nations groups, who are articulating their desires through GIS. In the United States, more than one hundred tribes are using some form of GIS. Land claims, in court, require spatial precision in order to be acknowledged by the court. Frank Duerden and Valerie Johnson, at the GIS '93 Symposium in Vancouver, B.C., commented: "GIS proves to be a useful tool in bridging the gap between traditional landscape images and the demand for formal cartographic representations of land necessary for land claim negotiation." GIS is being used for land claims in the Northwest Territories, Yukon, northern British Columbia, Alberta, and Saskatchewan. The Dene Mapping Project has overlaid government map sheets with "hunting and trapping trails; fishing areas; species sought; the years, seasons, and frequency of use; and in some cases cabins, camps, and important cultural sites."[41] The ultimate goal of these GIS claims—to prove former or present occupation—has already been attained in many cases.

In the 1970s, the Inuit and Cree in Quebec began mapping with GIS in order to negotiate with Hydro-Québec. The Inuit of the Northwest Territories, shortly afterward, began a Land Use and Occupancy Study that became the basis for the establishment of Nunavut, an Alaska-sized territory, which ultimately became Canada's largest land claim award, twenty years later.[42] Since then, a planning commission has been set up "to gather the sophisticated knowledge of the Inuit, who know every river and ridge, but also to use the latest computer mapping technology (GIS) to produce an infinitely malleable, easily updateable series of digitized maps on which to base its land use plans and ensure the long-term sustainability of Nunavut's land, waters, and wildlife."[43] Ironically, most of the maps of the North already come from Inuit knowledge, though the Inuit are not given credit for this authorship. Early explorers often asked the Inuit to draw maps of the surrounding country for them, and the Inuit produced maps up to 150,000 square miles in dimension. Inuit maps led, in fact, to a great boon in arctic travel and exploration.[44] But Inuit maps were often substantially different from European maps, leaving explorers at a loss for using the information. Though Inuit tended to have no problem understanding explorer maps, Inuit maps were often too sophisticated for Europeans. Inuit, for instance, often included "sky maps," stories, memories, and experience, along with drawings in sand or carvings on wood to represent the landscape. Because the arctic landscape is constantly changing, Inuit watch the sky for "water sky," which is where a pocket of water has opened in the sea ice, as witnessed by the darker sky over this area.

Barry Lopez describes a scene in which British explorers watched on the beach as three or four Inuit drew a map for them in the sand. Lopez writes: "They did not know what to leave out for the impatient men. . . . Therefore, as they placed a line of stones to represent a mountain range and drew in the trend of the coast, they included small, seemingly insignificant bays where it was especially good to hunt geese, or tapped a section of a river where the special requirements for sheefish spawning were present. This was the map as mnemonic device . . . three of four men unfolding their meaning and purpose as people before the young officers." The officers found the maps "exotic and engaging" but did not know how to read them. Robert McClure, in contrast, recorded a time when the Inuit drew an exacting map with pencil and paper "as if they were accustomed to hydrography."[45] It appears, in fact, that the Inuit developed maps as dialogue with explorers, sometimes attempting to share vital information that the explorers could not follow and other times understanding exactly what the explorers were looking for and, obligingly, producing it. Sunapignang of the Cumberland Sound–Frobisher Bay, for example, drew a map from memory that, when compared with a modern map, shows a remarkable similarity (see fig. 21). The main differences, in fact, are due to distortions in size for areas that are hunted more frequently or have more favorable fishing.

First Nations' GIS use in Canada today demonstrates that GIS no longer has to be—to use Taylor and Johnston's terms—(A) collecting information about (B). Instead, it has become a means for (B) to defend its territory and itself from (A) and thus to gain sovereignty. Nunavut, for example, plans to use GIS to clean up PCB contaminants left by the military in the 1950s, to monitor marine shipping routes to determine their effect on wildlife, to map the disruptive effect of aircraft and helicopter noise on wildlife, to map archaeological sites, and to provide maps with local Inuktitut names in order to preserve culture for the next generation. If early mapping projects, among the Inuit, often occurred in dialogue with explorers after contact, today indigenous peoples are continuing this dialogue by learning to tell their stories and make their demands in a way that the courts understand.

GIS is also allowing Native American groups to transgress or overcode arbitrarily imposed national boundaries, demonstrating the need for international land-management practices. The Aboriginal Mapping

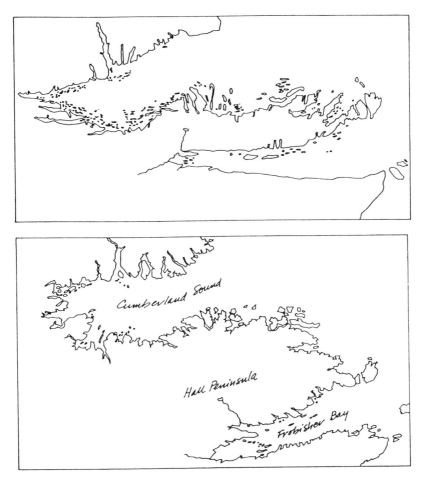

21. Above, map of Cumberland Sound–Frobisher Bay drawn from memory by Inuit Sunapignang; below, a modern map of the area. Reprinted from Sixth Annual Report of the Bureau of Ethnology, Smithsonian Institution, 1888.

Network, a Canadian network designed to share GIS information with First Nations, received an e-mail from Tom Curley, the GIS program manager for the Suquamish tribe in Washington, which read: "Feel free to e-mail information that might pertain to Tribal GIS down here—after all, the Canada/US boundary is artificially imposed."[46] This type of post-national movement is a far cry from its usual definition as "the graceful withering away of all nation-states."[47] Postnationalism, in the latter sense, is merely the ideological superstructure for the economic base of multinational corporations. It does not reflect the interests of First Nations, who reject both the nationalism of Quebec and the trans-nationalism of NAFTA. For the Cree, NAFTA means that hydroelectric power must continue to be supplied to the United States, regardless of the cultural implications for First Nations.

GIS was designed for the exploitative takeover of indigenous lands, the erasure of their history, and the occupation of the North. This became a way for Canada to manifest its sovereignty in areas that were "scarcely" occupied, thus denying native sovereignty. But like the Indians who instantly understood Susanna Moodie's map and relabeled it—in their own minds—with indigenous names, today with GIS the First Nations are taking state-produced maps and relabeling them to reflect indigenous history. Susanna Moodie may have tried to pull away the map, alarmed by the Chippewas' "strange, uncouth exclamations," but today First Nations are taking the maps back. This is not to imply that this struggle has not met with opposition. First Nations GIS development has been seen as threatening to other land demarcation companies, to the point that the organization representing surveyors in Canada has lobbied against the training of First Nations communities in GIS and GPS technology.[48] The struggle over who has the rights to map land continues, whether it is by U.S. satellite, Hydro-Québec, or Algonquin elders. Susanna Moodie and that "uncouth" Chippewa remain with their hands on the map, each understanding what control of it means.

J. J. Otero wrote on April 24, 1996, on NativeNet, an e-mail discussion group for American Indians: "You could be an Indian if you walk down the hall of a big corporation and someone asks you if you could mop up the mess that they made. And of course you oblige and afterwards come into their office and hook up their network connection to the mainframe via a tcp/ip protocol over a fiber backbone." Not surprisingly, this was never the type of Indian that Margaret Atwood seemed to want to be, nor did going native ever seem to imply hooking up a computer network connection or designing a GIS program. Atwood's desire to become native, "at least spiritually," is simply a continuation of "Grey Owl's longing for unity with the land, his wish to claim it as homeland."[49] The "at least" demonstrates that Atwood may not want to live the political reality of today's First Nations, or even to recognize the prior claim of the Algonquin to her hometown. She does not want to mop up the mess or fight for First Nations land claims, because Atwood's desire, in and of itself, is a land claim, a desire to "claim it as a homeland." David Spurr suggests that, in colonial discourse, "negation acts as a kind of provisional erasure, clearing a space for the expansion of colonial imagination and for the pursuit of desire." Spurr does not deal with the concept of going native or filling that void with an idealized native-white hybrid who displaces the real inhabitants. Instead, in Spurr's summation, going native is equated with "the supposed danger of the European's degeneration in the presence of the primitive," which leads to "obsessive reprehension" for the natives.[50] But in Atwood's scenario, there is neither reprehension nor idealization of the natives; there is, instead, idealization of herself as native, in her Grey Owl costume. To her credit, Atwood did not choose the Grey Owl garb; instead, she chose "my Baffin Island Inuit female skinning-knife earrings" for the lecture on the North, and a fringed leather jacket with "fringed leather earrings" for her Grey Owl lecture.[51] She said the British audience seemed especially entertained as she went on to explain the native myth (as represented in non-native literature) of the "cannibalistic" Wendigo, a man-eating creature that can infect Indians and turn them into cannibals as well. Maybe the British audience even recalled Kipling's tales a century earlier of the head-shrinking practices of Afghans or Stanley's lectures on cannibalism in the Congo. It would have been interesting to see, in comparison, if they would have responded so favorably to a presentation by a First Nations group declaring their own sovereignty through GIS maps.

CHAPTER 6

Postcolonial Occupations
of Cyberspace

On March 20, 1974, while sitting at home in Orange
County, Philip K. Dick was seized by VALIS, or the Vast Active Living
Intelligence System. He explained: "It seized me entirely, lifting me from
the limitations of the space-time matrix; it mastered me as, at the same
instant, I knew that the world around me was cardboard, a fake. . . . It
took on in battle, as a champion of all human spirits in thrall, every evil,
every Iron Imprisoning thing."[1] This hallucinatory experience, which
may have been precipitated by a mild stroke or an LSD flashback, would
come to dominate his writings until his death. In 1980, he wrote a book
called *VALIS,* in which Valis was a creature of pure information that had
been buried in the ground at Nag Hammadi. Dick defined Valis as "a
spontaneous self-monitoring negentropic vortex . . . tending progressively to subsume and incorporate its environment into arrangements of
information" with "quasi-consciousness, purpose, intelligence, growth
and an armillary coherence." It could be described as a bizarre cross
between artificial intelligence, virtual reality, and the Gnostic God. Valis,
in Dick's life, manifested itself through a series of revelations in which he
was struck by a "pink beam" of data, heard an "AI" voice speak to him
from outer space, was explained the secrets of time by a woman deliver-
ing pizza, or watched as "the landscape of California, U.S.A. 1974 ebbed
out and the landscape of Rome of the first century C.E. ebbed in." Valis,
for Dick, was both a book and a part of his daily life.[2]

"Only Phil could write an autobiography and have it be science fiction," Dick's agent once remarked. Robert Toth described his work:
"Imagine Franz Kafka writing 'Twilight Zone' episodes: Dick's characters,
more often than not, wake up to discover their world is a stage set constructed to keep them under close observation. Even more often, they
wake up to discover that they're robots—or aliens or part of a cult of

Gnostic Christians—and never realized it."[3] In Dick's "I Hope I Shall
Arrive Soon," a man cryogenically frozen on a trip to a colony on another
planet partially wakes up due to a malfunction in the freezing process during the ten-year trip. In order to keep the space traveler from becoming
psychotic, the computer monitoring the trip decides to feed his brain both
his own memories and computer-generated fantasies of his arrival at the
colony. When the passenger finally arrives at the new colony, he is unable
to believe that he is not being fed computer-generated fantasies that will
abruptly end at any moment and return him to his past. He is, in fact,
psychotically trapped within a computer-generated reality of himself.[4]

Dick's fear and fascination with the idea of being swallowed by a
false or computer-generated reality is combated, interestingly enough, by
another computer-like system, the Vast Active Living Intelligence System. Valis, rather than trapping the individual in a false reality, frees him
or her from "every Iron Imprisoning thing." In Dick's language, this
imprisoning thing is the institution of a corrupt state, whether it be
Rome or Hollywood or the Nixon administration. Dick's notion of the
imprisoning thing overlaps with his belief in the evil Gnostic God called
the Demiurgos (or "half-maker") because half of what he makes is considered divine and half is flawed. Erik Davis, in *Flame Wars,* writes: "For
Dick, the ancient demiurge is recast as the irrational 'Empire': Rome, the
Nixon administration, the State as such. Dick did not emphasize the
material or Satanic aspect of demiurgic powers, but rather their ability to
create false worlds."[5] It is an anxiety about a state-produced representation of the world.

Dick's "quasi-conscious" Valis, which tends to "subsume and incorporate its environment into arrangements of information," could be a
Gnostic-laden description of GIS. This seemingly agonistic combination
of religion and computers has been explored by other sci-fi writers, such
as Clive Barker and British writer Barrington Bayley. Barker recently
moved away from splatterpunk to Gnostic territory in his novel *Galilee.*
Bayley made the claim that "what the science fiction writers were doing
before there was any science fiction—they were writing the Gnostic
myths."[6] Dick, after his experience of Valis, wrote two novels dominated
by Gnostic themes, as well as a private exegesis of more than eight thousand pages on Gnosticism.[7] Amitav Ghosh, who later wrote *The Calcutta
Chromosome,* worked as an anthropologist in Upper Egypt when the

discovery of the Gnostic texts occurred in Nag Hammadi. Considered to be writings from the Gnostic cult of Valentinus, these buried texts were suppressed and hidden by early orthodox Christians intent on developing homogeneity within the religion.[8] Ghosh's *Calcutta Chromosome*—which recently won the Arthur C. Clarke Award for best science fiction novel—turns this moment of discovery into the basis for a counter-science sci-fi novel.

So what is the connection between Gnosticism and science fiction? Gnosticism's greatest foe, Ireneaus, may explain this connection most clearly in his description of the Gnostics: "Each one of them, as far as he is able, thinks up every day something more novel. None of them is perfect if he does not produce among them the greatest lies." And: "They adduce an untold multitude of apocryphal and spurious writings, which they have composed to bewilder foolish men and such as do not understand the letters of the Truth."[9] Gnosticism, in this sense, is a language of proliferation at a time when certain leaders in Christianity were severely attempting to limit the dogma and discourse of Christianity. If the Gospels were allowed to proliferate or "develop mighty fictions," or if more people were allowed to tell their version of the story, we might have ended up with a religion steeped in the doctrine of orality. Instead, the power struggle within early Christianity led to the selection of canonical texts, which involved the deification of texts deemed authentic and the burial of others deemed fictional.

Similarly, science fiction—at least Dick's version—resists the homogenous "truth" language of the state in favor of proliferating technologies that potentially subvert the state's authority. Postmodern theorists have suggested that cyberspace, as a decentralized mode of communication, represents an agonist to grand narratives, totalization, centralization, and industrialization.[10] Mark Nunes wrote: "Internet, rather than presenting a simulation of totality, might provide a space of play. Rather than pursuing ends through this technology, one might instead turn oneself over to the drift and *derive* of 'cyberspace.'"[11] Indeed, theorists of cyberfiction often focus on games: Adventure, SimCity, Dungeons and Dragons. Cyberfiction itself is often playful, as this description by Erik Davis demonstrates: "In the novella *True Names,* [Vernor] Vinge describes the Other Plane as a virtual representation of 'data space' accessed by game interfaces called Portals. . . . The hacker denizens of the Covens perform various pranks for fun and profit, and take on colorful handles like Mr. Slippery and Wiley J. Bastard; like D&D players, they construct the imagery of their characters, most choosing to represent themselves as magicians and witches."[12] Self-construction, trickery, game spaces, and role playing provide the pleasures in cyberpunk. Cyberspace, many contend, is a performative realm where initiates may author themselves.

Gnosticism, in science fiction, functions less as a religion than a metaphor for the possibility of a counter-religion, just as the technology is often of the subversive, makeshift, or hacker variety. It has the ability to invade and destroy the monolithic technology of the state or corporation. John Pickles claims that GIS similarly rejects "the foundational claims of positivism" and presents a "writerly text." Because GIS presents "multiply embedded systems of texts," Pickles claims, "all texts are . . . embedded within chains of signification: meaning is dialogic, polyphonic, and multivocal—open to, and demanding of us, a process of ceaseless contextualization and recontextualization."[13]

SIMULATED LOCALIZATION

GIS is actively building a new world on-line, as users remark: "We are only at the beginning of the process of delimiting and mapping the territory and content of this new currently solidifying *terra incognita.*"[14] Philip Dick's new world is Valis, which "subsume[s] and incorporate[s] its environment" into itself. In Amitav Ghosh's *Calcutta Chromosome,* AVA II is working on a global inventory for the International Water Council. Caught in the cataloguing fever of a new millennium, Ava is programmed to enter all the "endless detritus of twentieth-century officialdom—paper-clips, file-covers, diskettes." The International Water Council is an imperialistic organization that has slowly absorbed all other global agencies, and so the records of the old global organizations are being consolidated into Ava. Ghosh writes, "Instead of having a historian sift through their own dirt, looking for meanings, they wanted to do it themselves: they wanted to load their dirt with their own meanings."[15] The International Water Council wants to control its own narratives and write its own history, with no detail left out to be found and interpreted by others.

The International Water Council appears to function as a metaphor for the control of information. It seemingly has no real connection with water anymore, which has become a kind of archaism in the system. The flow of information is continually described by the metaphor of water, as it crosses "watersheds" around the world. The Water Council—more of an information council—is invested in controlling and monitoring these flows. The Water Council stores a map of the world, which contains details down to regional dialects that have long been forgotten by the general population. It is a recorder of history, remaking historical information to suit purposes that no one can understand.

Ava is located in Manhattan, in Antar's apartment; Antar works at home for the Water Council. Ava has the ability to "simulate 'localization,'" speaking to Antar in the dialect of his home in the Nile Delta (14). Ironically, Antar himself has lost the ability to speak his local dialect. Just as water has become a metaphor for something so controlled that it no longer really exists in the natural world but only in reservoirs, so the local is more perfectly simulated in the controlled environment of the computer. Like Baudrillard's hypertelia, where the sophistication of a model becomes greater than the reality it attempts to comprehend, Ava's reproduction of authentic dialects is replacing the spoken languages themselves.

Ava can project holographic images from around the globe, creating virtual realities that the user can "enter" for international "face-to-face" conversations. Though Antar's Manhattan apartment complex is becoming abandoned and replaced by "storage space," he can fill his room with people from Calcutta—where the Water Council headquarters are located. Antar's work entails maintaining his symbiotic relationship with Ava by answering the occasional question that Ava asks about an object that she has found. Ava, in turn, provides Antar with memories of his childhood home in Egypt, which Antar no longer really remembers.

The Calcutta Chromosome represents a new genre of postcolonial science fiction, inasmuch as it rewrites the history of colonial scientific discoveries while at the same time envisioning the possibility of a virtual homeland—or, that is, the impossibility of recreating a real home after it has been disrupted by colonialism. In *The Calcutta Chromosome,* Antar's entire village in Egypt has been killed by a rare strain of malaria introduced to the village by a Hungarian archaeologist. Home, in this sense, is symbolically an impossibility at the beginning of the novel. All that remains, in this futuristic novel, are simulations of home, vestiges of a forgotten past that are stored in virtual reality. Similarly, Salman Rushdie, in "At the Auction of the Ruby Slippers," creates a futuristic world in which movie characters can step off the screen and marry members of the audience. The narrator complains: "This permeation of the real world by the fictional is a symptom of the moral decay of our post-millennial culture. . . . There can be little doubt that a large majority of us opposes the free, unrestricted migration of imaginary beings into an already damaged reality, whose resources diminish by the day." The most important event in this postmodern world is the yearly auction, where national symbols are for sale: the Taj Mahal, the Statue of Liberty, the Alps, the Sphinx. The ruby slippers from *The Wizard of Oz,* however, are the most expensive item at the auction, because of their "affirmation of a lost sense of normalcy." The narrator hopes to buy the slippers for his lost love so he can whisper in her ear, "There's no place like home."[16]

G. Roulet has written that the postmodern "technical utopia" is based upon a "spatialization of communication" in which "all localiza-tion becomes impossible." The production of this utopia, he concludes, means "the final dissolution of all ties and places that symbolically struc-tured traditional society."[17] Ironically, however, these ties and places can be artificially resurrected by computer, as the agency of the subject becomes invested in these resurrections as fetishized objects of home or normalcy. Ava resurrects home, just as she maps the "currently solidify-ing *terra incognita*" of a global agency. Interestingly, Ava is produced in India, and the majority of global information flows through Lhasa because of its central location between the world's major "watersheds." Science fiction, in contrast, is generally set in one or more of the most high-tech or industrialized cities, such as Los Angeles or Tokyo. *The Cal-cutta Chromosome* begins in New York, but even New York is predomi-nantly settled by immigrants: Antar is from Egypt, and his neighbors are Indian. The novel then quickly moves back in time to Calcutta.

The Calcutta Chromosome represents the survival of a counterscience, or counterhistory, at the heart of colonial history. Indeed, the colonial version is revealed to be arbitrary and mistakenly imposed; it is the fabri-cation that the counterculture group allows to exist in order to conceal its covert actions. Ghosh's *In an Antique Land* similarly depicts a "small remnant" of an alternative Jewish-Muslim-Egyptian culture that survived "in defiance of the enforcers of History."[18] *In an Antique Land* charts the survival of hybrid affiliations in ancient history. *The Calcutta Chromosome* projects an alternative syncretism into the near future, drawing together Egypt, India, and New York City into one cabalistic revolution.

Gauri Viswanathan, in her article "Beyond Orientalism," claims that Ghosh's "engagement with the romance of syncretism . . . as a solution to sectarianism, nationalism, ethnocentrism, and religious intolerance, evokes a nostalgia that is itself unsettling." Discussing *In an Antique Land,* she writes: "No matter . . . how appealing his humanist call for dissolving barriers between nations, peoples, and communities on the grounds that world civilizations were syncretic long before the divisions introduced by the territorial boundaries of nation-states, the work cannot get beyond nostalgia to offer ways of dealing with what is, after all, an intractable political problem."[19] This common critique of syncretic values is certainly not to be underestimated; however, the type of syncretism that Ghosh presents is less nostalgic than subversive or counterhistorical. He deals in forgotten syncretic moments in history that have the potential, when unleashed, to change the reality of the present. Antar, for instance, survived the devastation of his village by malaria only to become the carrier of a disease that infects—at least in an allegorical sense—the whole world. This disease, in turn, becomes a virus that infects the computer-generated reality of the twenty-first century presented by Ava. Ghosh brings the virus back to New York City in an act of terrorism that only the "initiated" or infected can understand—he proposes the postcolonial occupation of the map. If this is nostalgia, then it is a dangerous form.

The Water Council, in *The Calcutta Chromosome,* could be read as a complex metaphor for the nationalism of India. In the same way, Salman Rushdie uses Saleem, a child born at midnight on Independence Day, as an allegorical figure for India in *Midnight's Children.* The year of India's independence is also, in *The Calcutta Chromosome,* the year that Antar was infected with the malaria virus and the Hungarian archeologist discovered the shrine of Valentinus, the Gnostic saint, in Upper Egypt. The history of the pseudo-Gnostic cult in this novel parallels that of India's independence, which parallels Antar's transformation. The secret that Antar must discover, in this novel, is his own history, which is also India's.

If the independence of India provides a sort of fulcrum in this novel, it should not be surprising that water is the metaphor of choice for the governmental authorities. "Dams are the Temples of Modern India," Nehru said in a speech after independence in 1947. Since then, there have been three thousand medium and large dam projects in India. "Dam-building grew to be equated with Nation-building," Arundhati Roy wrote, estimating that at least thirty-three million people have been displaced by these dams, primarily tribal people. She writes: "The ethnic 'otherness' of their victims takes some of the pressure off the Nation Builders. . . . The millions of displaced people don't exist anymore. When history is written they won't be in it. Not even as statistics." Controlling the flow of water is like controlling history. One billion people in the world live without safe drinking water, primarily in rural areas; yet these rural areas are being sacrificed, as villagers are forced to leave their land, so that more water can be transported to the cities. Roy writes: "India lives in her villages, we're told, in every other sanctimonious public speech. That's bullshit . . . India doesn't live in her villages. India dies in her villages. India gets kicked around in her villages. India lives in her cities. India's villages live only to serve her cities." Big dams supply water to urban areas because it is cost prohibitive to build the infrastructure necessary to transport water to more remote regions. In India, whole rural villages are threatening to drown themselves in the rising dam waters, rather than move. They know that, if they move, their living conditions will only deteriorate, even become untenable, in the resettlement communities. During the submergence of Narmada Valley, villagers stood in the water for twenty-eight hours, until the water reached their necks and they were dragged off by authorities. One seven-year-old girl, Lata Vasave, drowned when she got caught in quicksand-like silt. She had been living in a village that was gradually being flooded after the closing of the dam sluice gate. But the history of these people is submerged, along with their children. Morarji Desai, at a public meeting in the submergence zone, threatened: "We will request you to move from your houses after the dam comes up. If you move it will be good. Otherwise we shall release the waters and drown you all." Today there is a cement replica of the ancient Shoolpaneshwar temple on the shores of the reservoir; the real temple is under the water. At the Shoolpaneshwar Sanctuary Interpretation Center, the drowned villagers have also been resurrected. There is a life-sized thatched hut, with a dog asleep on the floor and food on the fire. Outside, there is a papier-mâché couple, smiling—"Mr. and Mrs.

Tribal." Arundhati Roy writes that, even in their replicas, "They're not permitted the grace of rage."[20]

What makes this vision possible is GIS, the World Bank, and India. Because of the complexity of managing such big dam projects—including drainage, irrigation, and groundwater—the government claims that a single authority must use a GIS water management program to electronically distribute the water and control the mix of fresh and salt water. Roy writes: "Who will sell the water? The Single Authority. Who will profit from the sales? The Single Authority. . . . Computer water, unlike ordinary water, is expensive. Only those who can afford it will get it. . . . The Single Authority, because it owns all the computer water, will also decide who will grow what."[21] Most of these big dam projects have been funded by the World Bank, which now requires a mandatory GIS component to development schemes. In the case of the Sardar Sarovar reservoir, the local is being simulated even as it is being submerged and made subject to the single authority.

Dam building, which has been considered nation building, is really about the centralization of authority and control of demographics. GIS is able to map a new future around the dam, complete with simulated locals, five-star lakeside hotels, and controlled agriculture. Nehru saw building a dam as a way to build a nation, but at the same time he accepted the conditions of the World Bank. He allowed a dependent relationship with first-world countries to emerge in order to support his development schemes. In reality, the building of big dams has always crossed national boundaries. The James Bay project in Quebec is for the purpose of supplying power to the United States. And today development plans for water use are being subsumed under an international single authority, the International Water Management Institute (IWMI). IWMI has developed a World Water Atlas, which provides "a computer-based atlas of world climate and related water and natural resources data in a format that could be quickly and easily used in GIS studies." The atlas has been sponsored by the Japanese government and the U.S. Agency for International Development and is available on the World Wide Web. Besides designing the GIS Water Atlas, the IWMI worked on water management projects in Gujarat, India, where the Sardar Sarovar reservoir now stands. The director commented: "The challenge for IWMI over the coming five years is to establish itself as The Global

Water Center. The fact that the Institute has the expertise and a reputation for scientific excellence in integrated water resources management makes it the ideal player to take this place."[22] This sounds much like the "Single Authority," as well as Amitav Ghosh's International Water Council.

Ghosh's futuristic novel captures the anxieties of a world that, as it is mapped, subsumes whole histories and peoples. The International Water Council is a metaphor for this control of information, which is a control of history. But it also expresses how anxieties about the scarcity of the world's water supply are leading to more and more centralized agencies for mapping water resources and making decisions about how water should be used. If water use decisions are made by an international single authority, history is truly controlled. In this sense, water is history. The kind of technology that will supply the map of this new world is GIS, or Ava. In return for relinquishing the control of water, Ava supplies simulated localization, preserving imaginary histories. In *The Calcutta Chromosome,* water stations are generally located outside of dry ditches, and the station managers are not sure what they are supposed to be doing there because the water is gone. The International Water Council has so consolidated and centralized water resources—most likely behind big dams—that the streams are dry, and the history of the landscape is either forgotten or submerged. The only requirement of the workers, then, is to follow Ava, without knowing why.

POSTCOLONIAL SCI-FI

In the dystopian environment of *The Calcutta Chromosome,* Ghosh injects the history of colonialism. If sci-fi generally occurs in the most industrialized nations, Ghosh collapses the distinction between first and third world. In this sense, he redresses the exclusion of "developing" countries from the discourse of sci-fi. That very exclusion functions as a kind of primitivism, which assumes that postcolonial nations are not the realm of aliens, cyborgs, and technocracies. Postcolonial nations are often described as being behind in development, or still catching up. In reality, GIS use is one of the fastest-growing industries in the so-called third world. Ghosh brings the reality of the postindustrial sector to Calcutta, creating an interesting blend of sci-fi and Gnosticism. As one critic described it: "Genetic engineering, precognition, shape-shifting, ancient

Egyptian mysticism and global mind control are combined here in a strange plot that's worthy of 'The X-Files.' But the narrative is also a game with narrative itself: Ghosh's characters are not so much telling the story as being told by it."[23]

The Calcutta Chromosome opens with a question from Ava. During her global inventory she has discovered an unidentified object, which Antar determines originated in Calcutta. The object turns out to be Murugan, who had disappeared in Calcutta while studying the history of Ronald Ross, who won the Nobel Prize in 1903 for reputedly discovering the cause of malaria. Antar had known Murugan a few years earlier, when they had worked together at an international health organization called LifeWatch, which was later absorbed into the International Water Council. Murugan had been obsessed with Ronald Ross's discovery of the malaria parasite. "Malaria was the cold fusion of his day," Murugan explains (55). Ronald Ross, a British colonialist in India, was on the trail of malaria simply because of the notoriety he thought it could bring him. If artificial intelligence is the dominant new technology in Antar's time, we find that malaria, virus theory, and genetic engineering were the cutting edge for Ronald Ross. Ross, like the International Water Council, was obsessed with recording the narrative of his discovery: "He wants everyone to know the story like he's going to tell it; he's not about to leave any of it up for grabs, not a single minute if he can help it" (52). Yet while this colonial control of narrativization is occurring on one level, at a different level the reader begins to realize that this control is entirely a myth—the true agency is elsewhere.

LifeWatch was absorbed into the International Water Council partly because all government agencies ultimately became subsumed under this "Single Authority." Disease control is also specifically linked to water issues, particularly in the case of malaria. In fact, one of the main objects of the IWMA is to control the spread of malaria through controlling water distribution. In the colonies, discovering a cure for malaria was of paramount concern because of the number of deaths that it caused among English colonists in unfamiliar, swampy environments. Both dam building and malaria control were directly linked to notions of colonial authority in British India. The Calcutta Chromosome, however, uses the illusion of control as a kind of subterfuge that allows the colonized to actually carry on their own discoveries. Murugan is obsessed with

Ronald Ross, it turns out, not because he discovered malaria but because ultimately he did not. *The Calcutta Chromosome* turns the whole notion of discovery on its head. Just as native guides actually led the explorers to their geographical "discoveries," Ghosh proposes that the familiarity of the Indians with malaria—their "intuitive understanding" of the disease—ultimately made it easier for them to find a cure.

Murugan finds that the servants of Ross's predecessor—D. D. Cunningham—actually made the discovery and handed it over to Ross. Mangala, D. D. Cunningham's servant, secretly established a laboratory in a shed and solved the problem herself. Her distance from the medical field enabled her to "think outside the box" and solve the puzzle. Murugan explains that "she wasn't hampered by the sort of stuff that might slow down someone who was conventionally trained" (246). But as the novel continues to unravel, we realize that this conquest of scientific history by "dhoolie-bearers" covers a more bizarre alternate history. Mangala found a strain of malaria that, when infecting a patient, can cause a crossover of personality traits from the malaria donor to the patient—via a pigeon. It is essentially a "technology for interpersonal transference" (107). And because people can continually transfer their personalities through the malaria virus, it is also the key to immortality.

The Gnostic leader Valentinus wrote: "What liberates us is the knowledge of who we were, what we became, where we were, whereinto we have been thrown, whereto we speed, wherefrom we are redeemed, what birth is and what rebirth."[24] In *The Calcutta Chromosome,* this spiritual cycle of rebirth is ultimately enacted in cyberspace—or whatever technology is available. At first, it is through the malaria virus; but in the end, the occult ritual of intrapersonal crossing occurs when Antar is brought into the Simultaneous Visualization program in Ava. Through an unexpected corruption of the rules of science, technology is used to proliferate the postcolonial voice and return mystery to the realm of science. Knowledge, in the novel, is also linked to mutation: "If it's true that to know something is to change it, then it follows that one way of changing something—of effecting a mutation, let's say—is to attempt to know it" (104–105). Mangala and her followers effect mutations in the system of knowledge, the malaria bug, and the group by revealing clues to those who become involved. The ultimate mutation in technology that these characters appear to be waiting for is the Internet. Murugan explains,

"Maybe they're waiting on a technology that'll make it easier and quicker to deliver their story to whoever they're keeping it for," which is Antar and Ava (219).[25]

When Ava finds Murugan's ID, a process is set in motion that ultimately leads to Antar's discovery that he himself is the answer to the mystery. Murugan explained to Antar: "They have to be very careful to pick the right time to turn the last page. See, for them, writing 'The End' to this story is the way they hope to trigger the quantum leap into the next. But for that to happen two things have to coincide precisely: the end credits have to come up at exactly the same instant that the story is revealed to whoever they're keeping it for" (218). In order to find the owner of the ID, Antar is given permission to open a direct line with the Water Council's representative in Calcutta. He sends a "raft" to the director's house and breaks in with a "shatter command" when he finds it locked, "entering" the house through a system of surveillance cameras. The director locates Murugan in an asylum in Calcutta, and when Antar demands that the director send for him, the director responds: "What if he experiences an alternative inner state while he's here? What if he wrecks the place? What if he wrecks the terminus?" (241) It is the echo of this threat that we are left with at the end of the novel, when Murugan holographically appears in Antar's flat—minus the director—and asks Antar to put on his Simultaneous Visualization headgear. Antar objects that the program is classified, but Murugan responds, "Guess we got in while the going was good" (310). After putting on the headgear, Antar immediately enters a room full of people who are saying, "We're with you; you're not alone; we'll help you across" (311). While critics have complained about the cryptic nature of this ending—indeed the ethics of the novel is cryptic knowledge—the experience of Antar entering the collective consciousness of cyberspace is much like the experience of Murugan being bitten by a mosquito and so biologically entering the cabalistic group.

In essence, what the novel achieves is a postcolonial takeover of cyberspace as well as a revision of colonialist forms of discourse and knowledge. The absolute implausibility of a malaria-infested, time-traveling cyberspace cult that has controlled the development of technological history is the novel's strength. Murugan is told by a cult member why he was not warned about the group's activities: "You would not have believed me. You would have laughed and said, 'These villagers, their heads are full of fantasies and superstitions.' Everyone knows that for city men like you such warnings always have the opposite effect" (281). *The Calcutta Chromosome* continually reverses expectations, causing "city men" to question the whole notion of superstitious knowledge.

In some ways, *The Calcutta Chromosome* resembles a bad *X-Files* episode on Satan-worshipping rural teenagers. By confusing the reader with images of candlelit cabal meetings complete with spurting blood, Ghosh disorients meaning within the text. By the twenty-first century, we can no longer distinguish who is who in the narrative—who has been infected by the virus/religion and who has not. By the end of the novel, we are not sure if this religious movement has taken over the International Water Council, or if the head of the Water Council was already one of them. In this bizarre counterhistorical, counterscientific novel, it appears that the colonized were actually always everywhere in control. Ghosh stages a postcolonial revolution in the occupied space of the Internet, creating a fascinating allegory of the failures of colonialism.

Ava, like Valis, subsumes the world, creating a new history according to its own logic. But Ava, like Valis, is also a religion, a revelation, and a Gnostic secret of divine knowledge. Both *VALIS* and *The Calcutta Chromosome* undermine the separation between science and religion. But while Valis is a metaphor for Philip Dick's schizophrenic breakdown, Ava represents the technology necessary for a postcolonial takeover of the world. Through Ava, history can be rewritten, as the fantasy of the future—told through GIS—finally falls into the hands of the third world. GIS is not intrinsically postmodern, as it can certainly represent the monolithic vision of the state. In the right hands, however, GIS can also represent the very serious business of postcolonial "play." It represents, in this sense, the staging of an organic revolution in the occupied maps of cyberspace.

Postmodern Primitivism

CARTOGRAPHY, for over a century, has sought to avoid "slogging into messy reality"—hoping to no longer need a world for its own confirmation. With the emergence of triangulation, aerial photography, and now space imaging, this goal appears closer to being realized. The need for "ground truth," what GIS practitioners call the fieldwork required to verify data, appears to be diminishing as data are instead checked against other data. The world is increasingly already stored within the computer or on the map. Satellite images of the world are now available at one-meter resolution, which is like "having a camera capable of looking from London to Paris and knowing where each object in the picture is to within the width of a car headlight."[1]

Al Gore, with his proposed Digital Earth project, hopes to consolidate these images of the globe into a three-dimensional representation of the planet that is available on-line. Gore suggested the benefits of this new earth:

> Imagine, for example, a young child going to a Digital Earth exhibit at a local museum. After donning a head-mounted display, she sees Earth as it appears from space. Using a data glove, she zooms in, using higher and higher levels of resolution, to see continents, then regions, countries, cities, and finally individual houses, trees, and other natural and man-made objects. Having found an area of the planet she is interested in exploring, she takes the equivalent of a "magic carpet ride" through a 3-D visualization of the terrain. . . . To prepare for her family's vacation to Yellowstone National Park, for example, she plans the perfect hike to the geysers, bison, and bighorn sheep that she has just read about. In fact, she can follow the trail visually from start to finish before she ever leaves the museum in her hometown.

So is the Digital Earth really for our children? Gore has said that, "working together, we can help solve many of the most pressing problems facing our society, inspiring our children to learn more about the world around them, and accelerate the growth of a multi-billion dollar industry."[2] The conflation of accelerating a multibillion-dollar industry and inspiring our children, however, reveals the kind of slippage that continually occurs in this type of rhetoric. A multibillion-dollar enterprise is rewritten as a family trip on a magic carpet that is smoothly operated for our pleasure.

In 1999, the world market for GIS products and services was estimated at $10 billion with a growth rate of 20–30 percent per year. The United Nations sponsored GIS projects in Sri Lanka, Bangladesh, Thailand, Jamaica, Liberia, Afghanistan, Kenya, Ghana, Pakistan, and Bhutan.[3] The World Bank made the adoption of GIS mandatory for all development projects in third-world countries. The GIS equipment was primarily manufactured in Europe or North America. Since the technical expertise was often not available to run these programs, first-world technicians were also often hired to train personnel to maintain the programs. Arturo Escobar, in response to these "development" projects, has suggested that the treatment of "developing" countries as objects of knowledge in contemporary discourse can only generate new "possibilities of power" in the first world.[4] As new resources are pinpointed though GIS, the possibilities of power are even further expanded.

Al Gore has also proposed, along with Digital Earth, a Digital Declaration of Interdependence, whose principles would include privatization of on-line services, open access to data, and universal service. Open access means the right to satellite (and other) data, which many countries have expressed concern over sharing. By pressuring institutions such as the World Bank, the Peace Corps, and USAID, Gore also hopes to see his dream realized of universal access to the Internet. The Peace Corps, he said, has promised to help increase access in telecommunications, and USAID has begun an initiative to promote Internet access and "electronic commerce" in eight developing countries.

Gore's new Digital Earth with a Digital Declaration of Interdependence mimics, obviously, the discovery of the new world with its eventual Declaration of Independence. Gore has actively promoted the use of GIS and Internet services in third-world countries, remarking: "We have a chance to extend knowledge and prosperity to our most isolated inner cities, to the barrios, the favelas, the colonias and our most remote rural villages . . . to strengthen democracy and freedom by putting it on-line, where it is so much harder for it to be suppressed or denied." Gore described an African farmer who would be better able to plant his crops because of access to a weather channel. He described a remote village near Chincehros, Peru, that had already "changed more in the past two years than in the previous century" because of the Internet. The village leaders, he claimed, formed an "on-line partnership with an international export company" and were able to ship their vegetables to New York City, raising their income from $300 to $1,500 a month.[5]

The question is whether this new earth will benefit the indigenous peoples of the world, as Gore promises it will. By putting global information on-line, some hope that economic development will occur; however, other critics suggest that the potential for development is "not supported and point to possible counter-productive effects, including an unreal, Disneyland-style democracy which only simulates reality." Gore's promise that democracy can be put on-line has been subject to intense criticism. In Mexico City, for instance, a project was designed to create on-line maps available to the public, including municipal boundaries and private properties, historical and ethnographic information, forestry regulations, and crop prices. Ricardo Gómez, Patrik Hunt, and Emmanuelle Lamoureux wrote that this project came, suspiciously, "at a time when millions of poor Mexicans are reeling from a programme of economic liberalization."[6] The project coordinator, Scott Robinson, suggested that "the Mexican telecenter initiative is only partially about computers and Internet connectivity. Its principal focus is on information policy—the availability and use of public domain information to . . . improve municipal administration and resource management, and create new opportunities for learning." Robinson claimed this project was "developing in tandem with a broader movement toward democratic reform."[7] If, indeed, one could say that "economic liberalization" (i.e., NAFTA) is beneficial for Mexico, then Gore's concept of Internet open access could be similarly described. However, as Mexico still "reels" from the effects of NAFTA, it is yet to be seen whether poverty rates will decline and profits will remain in the country.

The methodology of GIS may also be detrimental to women, as geographer Carol Hall has suggested: "The gains in status that women have made in geography are eroding because they do not utilize the technology to the extent that men do. In addition, GIS has added a masculine layer to the culture of academic geography. With the introduction of computer science (the basis of GIS technology), the masculine computer culture spread into geography more than it would have with less intensive use of the machines." Feminist geography, which emerged in the 1970s and 1980s, focuses on "how women have been hidden in much of traditional geographic analysis, thus leaving women's lives unexamined." GIS, however, which relies uncritically upon existing databases, tends to exclude women in the same way that it excludes indigenous peoples. Satellite images, which highlight large-scale or monolithic types of development, tend to obscure women's use of the land, from the backyard garden to the wildcrafting farm. Census reports, which have been critiqued for male-bias questions, similarly obfuscate women's role in culture and then are naturalized in GIS. Carol Hall writes, "Few existing databases contain as complete data about women or other marginalized groups as they do for dominant groups."[8] If the values of indigenous peoples and women are not included in GIS data, they will only be further marginalized in the culture.

The digitization of colonial-generated data in developing countries further problematizes the use of GIS. Peter J. Taylor and Ronald J. Johnston argue that "a data-led GIS geography would neglect most of the world. Analysis dependent upon rich data sources produces a rich countries' geography."[9] For instance, a GIS project in South Africa attempted to incorporate the "mental maps of the KaNgwane villagers" but found it largely impossible not to rely upon databases created under the former institution of apartheid. The project directors contended that any use of state-generated data in South Africa is likely to replicate the institution of apartheid, which relied heavily upon spatial data for the forced removals of blacks from "white" territories.[10] But the incorporation of KaNgwane mental maps became nearly impossible due to the imprecise nature of these constructs. "Cognitive information is geographically imprecise and is not comfortably expressed within a point/line/polygon paradigm," they explained. "This emphasis on fuzziness also runs counter to the GIS focus on quantification and error minimization." They ultimately concluded that community knowledge could only be transferred into a GIS if it was filtered by outsiders—and these outsiders would then replicate the very conditions of colonialism.

India similarly discovered the impossibility of evading colonial-generated data in its first national wasteland-mapping project using satellite data. Implemented in 1985 by Prime Minister Rajiv Gandhi, the project recorded those lands that had been designated as wastelands during British rule because they did not yield revenue to the imperial administration.[11] The project created a standard classification scheme for the entire country, defining wastelands as "land which can be brought under vegetative cover with reasonable effort, and which is currently under-utilised."[12] The wasteland-mapping project, however, did not take into account local or unsanctioned land uses or the role of the common lands in meeting the subsistence needs of the poor.[13] Indeed, its very purpose appeared to be to develop these lands in state-sanctioned ways, thus perpetuating the land use patterns established under colonial rule. Indian ecologist Vandana Shiva explained the problems behind this program: "The wasteland development programme . . . will in fact, privatise the commons, accentuate rural poverty and increase ecological instability. . . . The usurpation of the commons which began with the British will reach its final limit with the wasteland development programme as is."[14]

The continued use of colonial-generated data becomes a way to maintain the status quo and reinscribe indigenous or non-state-sanctioned lands as "barren," "wasteland," or simply "blank." GIS, when utilized for development purposes, will often focus on these barren lands and look for a way to extract resources from them, whether through irrigation projects, mineral exploration, or tourism. In areas where there is a lack of good or available data, satellites and remote sensing help to fill in the blanks. Because corporations or states create these data, rather than the local people, the area tends to be developed through their filters. In a GIS software ad, this connection between data and colonization is made obvious. The advertisement promises that you can "Govern your data" with this software (fig. 22). The software promises that it "centralizes mission critical information" and is built on a "rock-solid property data model"—the Govern logo suggests a queen's orb or a presidential seal. The message of the ad is that your data can be marked by you as "rock-solid property" that will always be available and pliable for your "mission." ArcView GIS advertises that you can go "into the heart of Africa" with their program. "The existence of ArcView GIS

22. Govern Software promises to keep potentially unruly, or ungovernable, global data under control. "Govern your data," 1998, courtesy of Govern Software.

allows our petroleum land management customers to tap the full power and true complexity of the data for the first time, both on the desktop and in the field," one corporate customer claimed. The "heart of darkness" is illuminated by GIS, revealing oil reserves, old-growth forests, or mineral deposits. In another advertisement headed, "Secrets of the Earth: genuine treasures for those who can exploit them," a satellite image of the globe is overlain with images of its resources: agriculture, coastal zone, environment, urban planning, and civil engineering (fig. 23). The ad implies that those who buy the product, Matra GIS, will be able to exploit these resources, which are waiting to be revealed. The caption at the bottom of the ad reads: "When the Earth reveals its secrets . . ." A recurrent trope in GIS ads today is that the earth is *waiting* to be developed by those who have the power to *see* its resources with GIS.

23. "We must, if needs be, place Nature on the rack and force Her to yield up Her secrets," wrote Francis Bacon. This software enables "direct and efficient exploitation" after the earth "reveals its secrets." "Secrets of the Earth," 1997, courtesy of Matra Systèmes and Information.

Developmental Primitives

The concept of the "primitive" has today been replaced, in the political world, with the "underdeveloped." Ivan Illich has said, "It took twenty years for two billion people to define themselves as underdeveloped."[15] In the same way that colonial powers were once attempting to civilize the colonies—by bringing them railroads, dams, and the Bible—so today nongovernmental organizations, the World Bank, and the U.S. Agency of International Development are trying to develop these countries by providing them with GIS, dams, and humanitarian ideologies. Interestingly, as "primitive" spaces are "developed" for resources, they are also artificially resurrected or preserved by the same GIS programs. An ad for a GIS program called the World Construction Set, for instance, depicts an idyllic Chinese village frozen in time with the help of GIS (fig. 24). With the "Rules of Nature" feature, the user can "instantly cover the terrain with realistic foliage in natural-looking, multi-level ecosystems." The perpetual resurrection of the past, the natural, and the "traditional" serves to counteract the anxiety of a new and invasive form of technology. The viewer is made to believe that the computer can bring back, rather than destroy, China or Bermuda. The owners of the product, the advertisements imply, have the power to make any kind of world that they want simply by choosing what to map and "visualizing" this selective new reality.

The belief that indigenous cultures should develop—or become modernized—and yet remain frozen in time depends upon severing the indigenous subject into two conflicting categories. Because indigenous societies were represented as precursors of Western civilization, preserving these cultures takes on the same meaning as preserving one's own past. Marianna Torgovnick explains that "primitivism" is based in the "assumption that primitive societies exist in an eternal present which mirrors the past of Western civilization."[16] Beyond this, we might say that primitive societies must be kept in an eternal present at the risk of losing our own history. And so indigenous peoples are only interesting as us and forfeit our interest when they become themselves. Territorially speaking, this manufacturing of the primitive as us allows the land to be cleared of primitive organization, or developed, while the primitive "culture" simultaneously becomes despatialized and incorporated into the democratic culture of the West. The separation of indigenous culture from the territory and the appropriation of the indigenous fetish into

24. The World Construction Set promises a more "authentic" landscape. World Construction Set, 1998, courtesy of Questar Productions, LCC.

capitalism serves to doubly disavow actually existing indigenous peoples. This dynamic has not changed with the emergence of so-called post-modernism yet remains largely unexamined.

The fascination with primitivism in contemporary Euro-American cultures has recently seen a phenomenal resurgence. In 1989, the publication of V. Vale's *Modern Primitives* fueled a new interest in what she calls the "neo-fetishist" practices of tattooing, multiple piercing, and scarification. "Obviously, it is impossible to return to an authentic 'primitive' society," Vale writes. "Those such as the Tasaday in the Philippines and the Dayaks in Borneo are irrevocably contaminated. . . . Under scrutiny many 'primitive' societies reveal forms of repression and coercion (such as the Yanoamo, who ritually bash each other's heads in, and African groups who practice clitoridectomy—removal of the clitoris) which would be unbearable to emancipated individuals of today." Instead, Vale claims, "modern primitives" are looking for a new, "more ideal society." The link between tattooing and an ideal society may appear dubious, but modern primitives believe that the transformation of the self—through pain—can "provoke change" in the social world. "Individuals are changing what they do have power over," Vale writes, "their own bodies." Pain is, according to Vale, the only way to prove one's own "authentic" and unique sense of self; the proof of existence is expressed through self-mutilation.[17] Vale's *Modern Primitives,* a text in mourning for the "authentic" subject, resurrects an ambivalent relationship toward indigenous peoples. Because we cannot reclaim authentic primitive societies, we will recreate ourselves as authentic objects.

Modern Primitives documented what has since become a vast subculture of fringe groups who call themselves "postmodern primitives," "urban primitives," "neoprimitives," or "anarcho-primitives." These various primitivisms are not limited to body modification practices but include raves, rituals, cyberpunk, paganism, tribalism, shamanism, and the celebration of solstices and equinoxes. This form of primitivism is a scarcely defined but thriving underground culture in many urban centers of Canada, the United States, Europe, and Australia. It is the drop-out culture of the sixties redefined as both indigenous and postmodern. Similarly, the *Fifth Estate,* a Detroit-based anarchist journal, described "an emerging synthesis of postmodern anarchy and the primitive . . . Earth-based ecstatic vision." In 1986, the *Fifth Estate* explained that they supported "a more critical analysis of the technological structure of western civilization, combined with a reappraisal of the indigenous world and the character of primitive and original communities. In this sense we are primitivists."[18] The publishing of Fredy Perlman's *Against History, Against Leviathan* in 1983 defined anarcho-primitivism for perhaps the first time.[19] John Moore writes that "the primitive, for those trapped in civilization, is a process, a process of renewal and return . . . a return to roots, but 'our' roots as they are now, in all their presence and sense of possibility, rather than some impossible search for origins."[20] The impossible quest for roots without origins becomes the dilemma of urban primitivists, who often simply make up conjectured indigenous rituals. Primitivism today is not focused on the issues of existing indigenous peoples but on the postmodern recreation of their imagined spiritual existence. In attempting to reclaim a romanticized indigenous past while ignoring or actively destroying their present and future, many Euro-Americans participate in a culture of primitivism. Primitivism is projected onto the territory as a site without good government or organization; as primitive societies are thus despatialized, they are nostalgically resurrected in the space of the individual.

It is not only in grassroots counterculture movements that a revival of primitivism is underway. Postmodern theory—based in the work of Nietzsche, Marx, and Freud—perpetually replicates the tenets of primitivism. Jean Baudrillard conflates the revolutionary and the primitive, as Steven Best and Douglas Kellner explain: "[Baudrillard is] dreaming of . . . a return to symbolic societies as his revolutionary alternative."[21] Jean-François Lyotard similarly claims that Baudrillard recreates "ethnology's good savage" in *Symbolic Exchange and Death;* ironically, Lyotard himself calls for a return to "paganism" in *Just Gaming* and "Lessons in Paganism."[22] This is not to suggest that Lyotard's pagans remotely resemble Baudrillard's primitives, but merely to point out the extent to which the trope of premodern reasoning is used in postmodern theory. Relying upon Marx's notion of "use value" and Marcel Mauss's use of the indigenous "potlatch," Baudrillard builds his system of "symbolic exchange"—a precapitalistic system of gift giving.

Gilles Deleuze and Félix Guattari similarly reinvest primitives with revolutionary authority, idealizing primitive territoriality as "a praxis, a politics, a strategy of alliances and filiations." They write, "In the primitive socius desire is not yet trapped, not yet introduced into a set of impasses, the flows have lost none of their polyvocity." Desire becomes trapped, they suggest, only in the nuclear, bourgeois family system that exists within capitalism. "Schizoanalysis," then, is a freeing of the flows that existed in the primitive socius—as opposed to psychoanalysis, which is based upon the "mommy-daddy-me" family structure. "Primitive cures" are "schizoanalysis in action," they explain, conflating the primitive and the "schizo."[23] Deleuze and Guattari also cite anthropologist Edmund Carpenter's work on Eskimo culture in *A Thousand Plateaus*. Carpenter writes: "Entire Eskimo societies are implosive: everybody is involved with everybody simultaneously and instantaneously. There is no isolating 'individualism' and no emphasis upon isolation of sight from other senses."[24] The "schizo," we ultimately must conclude, is the "Eskimo."

Primitives have been viewed, from Marx to neofetishists to Baudrillard, as models for regaining access to a body that has been recoded by the state, capitalism, and the nuclear family. Primitivism is nostalgia for a body that is not cyborg and a home that is not bourgeois. Torgovnick has written: "Going primitive is trying to 'go home' to a place that feels comfortable and balanced, where full acceptance comes freely and easily. . . . Whatever form the primitive's hominess takes, its strangeness salves our estrangement from ourselves and our culture."[25] But because the primitive can never be home, and is in fact constituted as the opposite of home, ambivalence emerges. Primitivism is a kind of ambivalent envy, in which indigenous persons are tacitly evaded while their practices are taken. The ambivalence, in this context, comes from the possibility that the indigenous persons might not want their culture and practices stolen or might even stand in the way. The Inuit may not want to be the representation of pretechnological society, either "Eskimo" or "schizo."

POSTCOLONIAL CARTOGRAPHY

This artificial "Disneyland-style democracy" of natives who are fetishistically frozen in time while economically forced to develop will continue unless indigenous peoples are allowed to narrate their own experiences through GIS. Cartography, in effect, is a struggle over narrative. Postcolonial critics have often linked geography to imperial domination, and with good reason. One need only look at Admiral La Roucière–Le Noury's opening address to the Second International Congress of Geographical Sciences in 1875 to gain this impression: "Gentlemen, Providence has dictated to us the obligation of knowing the earth and making the conquest of it. This supreme command is one of the imperious duties inscribed on our intelligences and on our activities. Geography, that science which inspires such beautiful devotedness and in whose name so many victims have been sacrificed, has become the philosophy of the earth."[26] Gore's mandate to "strengthen democracy and freedom by putting it on-line" resonates with the same utopian possibilities as "geography is the philosophy of the earth." By linking U.S.-generated ideologies to a new Digital Earth, Gore demonstrates this same expansive vision. What both of these narratives reveal is the way in which colonial and developmental rhetoric has animated the history of cartography. Because cartography is based in the perpetual superseding of "inaccurate" maps by "accurate" maps, geographic knowledge is believed to be progressive in nature. On the contrary, cartography may exist only as a perpetual digression from local knowledge in which the very object of the map is to destroy and replace local or indigenous knowledge. Cartography is based on an ambivalent and covetous relationship to indigenous knowledge, perpetuated by a colonial system that relied upon it and then erased it. In place of indigenous knowledge, the concept of the primitive emerged, which was considered the antithesis to modern knowledge.

In contrast, a nonexclusive narrativizing of the map could occur, in which indigenous histories are read as a part of the map. Through intentional processes of eliminating indigenous information—through distance from the native, the suppression of native names, the dismissal of the native guide's contributions, and the false projection of territorial ignorance onto the native—colonial domination occurs. In reality, colonial maps have been authored by indigenous groups who supplied information, kept explorers alive in the wilderness, and often literally carried them across the landscape. The trust of indigenous peoples was broken when natives were excluded from this historical process and their authority was stripped from the map.

Wilson Harris, who worked as a Guyanese surveyor for the British, described his experience in the fantastical novel *The Secret Ladder*. This semi-autobiographical novel centers on a Guyanese government surveyor who is sent to map a wild river but is constantly confronted with inconsistencies between aerial photographs and the view from the ground. Ultimately, he realizes that the river cannot be mapped with scientific scales, and the novel becomes a kind of allegory for postcolonial resistance against the British. Postcolonial nations are increasingly discovering these kinds of discrepancies between their land and the colonial map. In Guyana, for instance, the Amerindian population used GIS and GPS to map their own territories, which had been under dispute since independence. The government had issued a so-called Blue Map in 1982, which was supposed to delineate the boundaries of Amerindian villages. In response, the Amerindian populations, with assistance from the World Rainforest Movement, began mapping their lands in 1994. What the new maps revealed was that "the ambiguous Blue Map of 1982 was a wildly inaccurate relic of the colonial era, one that belonged in an antique shop." According to mapping specialist Peter Poole, "Rivers were misnamed, sources and connections were misidentified, entire watersheds flowed in other directions."[27] The Amerindian people have since been able to use their own maps to stop the development of a World Bank project, which could not occur on disputed Amerindian lands. Rather than the government receiving World Bank funds to establish national parks in the Amazon, the Amerindians are now asserting through their own GIS maps that ownership and control of these national parks should be passed over to them.

These types of mapping projects are not unique to the Amerindians of Guyana. In Nicaragua, Miskito divers are mapping their underwater territory with GPS. In Venezuala, the Ye'kuana are discovering errors of up to two kilometers by mapping their own territories with hand-held GPS units and GIS. Yakima (U.S.) and Innu (Canada) foresters have used GIS to create sustainable, locally managed forestry plans as an alternative to industry-based clear-cutting. In Chincha, Peru, fifteen indigenous mapping groups from Central and South America gathered to share mapping strategies. In Canada, the Gitxsan and Ahousaht First Nations, in cooperation with Ecotrust, created the Aboriginal Mapping Network. The work of groups such as these, especially in developing alternative resource-management plans, has led some nonprofit and management groups to propose the inclusion of indigenous input in regional and national planning. The Ryerson School of Applied Geography started an Indigenous Land-Use Information Project. The program coordinators, Frank Duerden and Richard Kuhn, explained the historical significance of this project:

Increasing attention is being paid to means of incorporating indigenous land use information in the overall land use and resource development process. . . . Pursuant to contact such information was widely used by traders and explorers to get to know the land. Over time, however, it was displaced by European-based knowledge, but interest in its use was revived over the past twenty years as debate over northern megaproject proposals grew and as the pace of native land claim negotiations accelerated. . . . Indigenous interpretations of landscape and environment were once more seen to have value and relevance.[28]

This project would help rectify the exclusion of native information from both the history and object of the map. Of course, for the state, this re-inclusion could ultimately represent a threat to both its cartographic industries and its control over definitions of the landscape. Peter Poole explains that "to project the power inherent in maps indefinitely, monopoly control is extended over the production and circulation of authorized cartographic versions."[29] These authorized versions are now being challenged by indigenous mapping projects that record their own histories of mapping—passed down in songs, sketch maps, and visual arts—in a GIS database. These mapping projects are largely centered around the "map biography," which includes such information as: "It was at this place that my daughter was born; or this is where my brother-in-law killed two caribou the winter a bear killed all my dogs; or this, Titiralik, is the place my snow machine broke down and I had to walk; Seenasaluq, this is a place my family has camped since before I was born."[30] These sorts of maps intrinsically link people to place, rather than severing them for purposes of establishing "objectivity."

In 1997, the Supreme Court in Canada made a landmark decision for aboriginal rights by placing oral histories on the same footing with written documentation in a court of law. First Nations in Canada are now mapping local knowledge at an unprecedented rate, "generating visual story boards of oral histories." While these maps can now be considered evidence in a court of law for land claims, there is a strong fear that the information revealed in these maps will be used against the First Nations. For instance, a GIS technician who designed forestry maps with the Innu explained that Innu maps that he was given to work with were provided only in strict confidentiality; he was not allowed to share them, or the information they contained, with anyone or keep copies after he was finished with the project. Ed Logan, who designed cultural resource maps for the Kwakiutl on Vancouver Island, intentionally generalized information in order to protect Kwakiutl interests. He said, "We take point data such as clam bed locations and represent it as generalized polygons, without a database." First Nations are concerned that they may be forced to reveal information collected in a GIS database through the Freedom of Information and Privacy Act. "First Nations are trusting the very agencies with which there has been a long record of conflict," Caron Olive and David Carruthers explained. Some indigenous groups fear that recording their stories in order to defend their version of history may only further the appropriation of their resources and lands.[31] Still, the Supreme Court decision in Canada to recognize oral histories marks a return of the aboriginal narrative to the map. If these narratives are allowed to proliferate, even though they may contest or overcode archaic colonial and resource-oriented maps, indigenous information may be reincorporated into planning and management as well as granted—albeit retroactively—the credibility that is its due.

On the other hand, if state and corporate data continue to be embedded in cyborglike technologies for first-world consumption, the history of the map may be further effaced. Tina Cary, president of the American Society of Photogrammetry and Remote Sensing, complained of current GIS technology: "My belief is that, to most consumers, the complexity of the technology is not yet sufficiently hidden. The complexity of an automobile or a fax machine or an ATM machine, in contrast, is hidden well enough."[32] Today, we have entered the era of cyborg production, hoping to finally hide the map in the body; but the more the map becomes embedded, the more it effaces its own racial constructedness and imperial origins. Jan-Willem Broekhuysen suggested that the world has become "an electronic 'global village,' in which neither time nor distance are barriers to the tide of information which now dominates our lives." Though this construction of the world is commonly seen as evidence of increasing decentralization, Broekhuysen continued: "That tide, like the sea itself, still largely ebbs and flows to . . . the beat which the Prime Meridian of Longitude at Greenwich has set on our world time system since 1884.[33] How many are aware that this "tide of information" is centered on Greenwich? This centering in effect is evidence that the imperial origins for the division of time and space have not been overcome but merely institutionalized to the point that they are invisible. It is because they are second nature to us (or cyborg) that they are able to function so effectively. As these divisions are embedded into user-friendly computer programs, they become even more invisible.

Anglo-identity has long been tied to the struggle over controlling territory, though this narrative has been strangely erased from the map, along with the indigenous subject, in order to naturalize the illusion of the map as objective. Europeans, in establishing sovereignty over a place, also believed they created themselves as sovereign subjects. In reality, indigenous people are the ones who discovered the discoverers, led them to food and water, and shared their territorial knowledge—only to have it betrayed by the final product, the colonial map. In reality, the person holding the globe in the GIS ads should be aboriginal.

Notes

INTRODUCTION   CARTOGRAPHIC CYBORGS

1. M. E. Clynes and N. S. Kline, "Cyborgs and Space," in *The Cyborg Handbook,* ed. Chris Hables Gray (New York: Routledge, 1995), 31.
2. For an overview of the use of images of the globe in GIS advertising, see Susan M. Roberts and Richard H. Schein, "Earth Shattering: Global Imagery and GIS," in *Ground Truth: The Social Implications of Geographic Information Systems,* ed. John Pickles (New York: Guilford, 1995), 171–195.
3. Joseba Gabilondo, "Postcolonial Cyborgs: Subjectivity in the Age of Cybernetic Reproduction," in *The Cyborg Handbook,* 424.
4. Steven Best and Douglas Kellner, *Postmodern Theory: Critical Interrogations* (New York: Guilford, 1991), 42.
5. S.F.H. Borley, *A Review of the Literature on the Use of Geographic Systems in Developing Countries* (Sheffield, U.K.: University of Sheffield, Department of Information Studies, 1991), 77.
6. M. R. Curry, "Geographic Information Systems and the Inevitability of Ethical Inconsistency," in *Ground Truth,* 79.
7. Quoted in Jean Baudrillard, *Simulations,* trans. Paul Foss, Paul Patton, and Philip Beitchman (New York: Semiotext(e), 1983), 2.
8. Quoted in Anne McClintock, *Imperial Leather: Race, Gender, and Sexuality in the Imperial Context* (New York: Routledge, 1995), 27.
9. Quoted in Barry Lopez, *Arctic Dreams: Imagination and Desire in a Northern Landscape* (New York: Scribner's, 1986), 282.
10. McClintock, *Imperial Leather,* 27–28.
11. Michel Foucault, *The Archaeology of Knowledge and the Discourse of Language,* trans. A. M. Sheridan Smith (New York: Pantheon, 1972), 228.
12. Jean-Paul Sartre, preface to *The Wretched of the Earth,* by Frantz Fanon, trans. Constance Farrington (New York: Grove, 1986), 26.
13. Mary Louise Pratt, *Imperial Eyes: Travel Writing and Transculturation* (New York: Routledge, 1992), 201.
14. Quoted in David Spurr, *The Rhetoric of Empire: Colonial Discourse in Journalism, Travel Writing, and Imperial Administration* (Durham: Duke University Press, 1993), 30.
15. K. Mayall, G. B. Hall, and T. Seebohm, "Integrating GIS and CAD to Visualize Landscape Change," *GIS World,* September 1994, 46–49. See also John Edward Fels, "Viewshed Simulation and Analysis: An Interactive Approach," *GIS World,* July 1992, 54–59.
16. J. Gottmann, *The Significance of Territory* (Charlottesville: University Press of Virginia, 1973), 17.
17. R. Y. Jennings claimed that international law historically "serves to divide between nations the space upon which human activities are employed, in

order to assure them at all points the minimum of protection of which international law is the guardian." Its purpose, more specifically, is "the delimita-
tion of the exercise of sovereign power on a territorial basis." R. Y. Jennings,
The Acquisition of Territory in International Law (Manchester, UK: Manchester
University Press, 1963), 2; "The Island of Palmas," Scott, Hague Court
Reports 2d 83 (1932) (Perm. Ct. 4rb. 1928), 2 U.N. Rep. Intl. 4rb. Awards
829; see also Sharon Korman, *The Right of Conquest: The Acquisition of Territory
by Force in International Law and Practice* (Oxford: Oxford University Press,
1996).

18. "The Island of Palmas."
19. This definition is from the CCH *Macquarie Concise Dictionary of Modern Law*
    (Sydney, Aus.: CCH, 1988), 129. In 1992, this definition was disputed and
    overturned in *Mabo & Others v The State of Queensland* (No. 2) (1992) 175
    CLR 1. 4. See M. Crommelin's *Law Institute Journal* 67, 9 (1993): 809.
20. See ibid.: "effective occupation is the traditional mode of extending sover-
    eignty over terra nulliua."
21. Robert. D. Sack, "Territorial Bases of Power," in *Political Studies from Spatial
    Perspectives,* ed. A. D. Burnett and P. J. Taylor (New York: Wiley, 1981), 67.
22. Ibid.
23. Daniel Defoe, *Robinson Crusoe,* New York: Bantam, 1981, 115.
24. Jean-Jacques Rousseau, "The Social Contract," trans. Gerard Hopkins, in
    Social Contract: Essays by Locke, Hume and Rousseau, ed. Ernest Barker (Lon-
    don: Oxford University Press, 1960), 187.
25. John Locke, *The Second Treatise of Government,* ed. Thomas P. Peardon (New
    York: Liberal Arts Press, 1952), 17
26. T. Carlos Jacques, "From Savages and Barbarians to Primitives: Africa, Social
    Typologies, and History in Eighteenth-Century French Philosophy." *History
    and Theory* 36, 2 (May 1997): 190–215.
27. Hayden White, *Tropics of Discourse: Essays in Cultural Criticism* (Baltimore:
    Johns Hopkins University Press, 1978).
28. Ibid., 180.
29. Max Horkheimer and Theodor W. Adorno, *Dialectic of Enlightenment,* trans.
    John Cumming (New York: Seabury, 1972), 3.
30. Asha Varadhajaran would later suggest that this concept of "nonidentity" is a
    useful tool for postcolonial theorists in reading resistance to the colonial proj-
    ect. However, given the complex "primitivism" associated with this project,
    this may prove to be a problematic task. Asha Varadharajan, *Exotic Parodies:
    Subjectivity in Adorno, Said, and Spivak* (Minneapolis: University of Minnesota
    Press, 1995).
31. Horkheimer and Adorno, *Dialectic of Enlightenment,* 31.
32. Sigmund Freud, "The Uncanny," trans. Alix Strachey, in *Collected Papers: Papers
    on Metapsychology: Papers on Applied Psycho-Analysis* (London: Hogarth, 1934),
    4:394, 403. Freud writes, "Each one of us has been through a phase of indi-
    vidual development corresponding to that animistic stage in primitive men"
    (394).
33. Ibid., 376.
34. Ibid., 393–394.
35. Ibid., 402.
36. J. Keay, *Explorers of the Western Himalayas: 1820–1895* (London: John Murray,
    1996), 188.

37. "A Memoir on the Geography of the North-Eastern part of Asia, and on the Question Whether Asia and America Are Contiguous, or Are Separated by the Sea," *Quarterly Review* 18 (May 1818): 434–435.
38. Clements R. Markham, *A Memoir on the Indian Surveys* (London: Allen, 1878), 125.
39. Patrick H. McHaffie, "Manufacturing Metaphors: Public Cartography, the Market, and Democracy," in *Ground Truth,* 119–120.
40. Clarence Winchester and F. L. Wills, *Aerial Photography: A Comprehensive Survey of its Practice and Development* (London: Chapman and Hall, 1928), 208.
41. This is from a paper by Aisha Ginwalla from my "Postcolonial Literature" course at the University of Missouri–Columbia (S2000).
42. McHaffie, "Manufacturing Metaphors," 116.
43. Timothy Findley, *Headhunter* (Toronto: HarperPerennial, 1993), 3.
44. Michael Ondaatje, *The English Patient* (New York: Vintage, 1992), 133.

CHAPTER 1   THE SUFFRAGETTE DISTURBANCES
              AND THE BOMBING OF GREENWICH

1. Christabel Pankhurst, *Unshackled: The Story of How We Won the Vote* (London: Hutchinson, 1959), 52.
2. Millicent Garnett Fawcett, *What I Remember* (Westport, Conn.: Hyperion, 1976), 181.
3. See Brian Harrison, *Separate Spheres: The Opposition to Women's Suffrage in Britain* (New York: Holmes and Meier, 1978).
4. Ibid., 43.
5. "Closing Days of the Trial [May 17–22, 1912," in *Speeches and Trials of the Militant Suffragettes: The Women's Social and Political Union,* ed. Cheryl R. Jorgensen-Earp (Madison, N.J.: Fairleigh Dickinson University Press, 1999), 205, 204.
6. Anne McClintock, *Imperial Leather: Race, Gender, and Sexuality in the Imperial Contest* (New York: Routledge, 1995), 46, 45.
7. George Egerton, *Keynotes* (London: Elkin Mathews and John Lane, 1893), 22.
8. Christabel Pankhurst, "The Political Outlook," in *Speeches and Trials,* 93, 94.
9. Christabel Pankhurst, "We Revert to a State of War," in *Speeches and Trials,* 127.
10. McClintock, *Imperial Leather,* 35.
11. Trevor Lloyd, *Suffragettes International: The World-Wide Campaign for Women's Rights* (London: American Heritage, 1971), 49.
12. Fawcett, *What I Remember,* 184.
13. Quoted in David Rubinstein, *Before the Suffragettes: Women's Emancipation in the 1890s* (New York: St. Martin's, 1986), 18.
14. Quoted in D. Howse, *Greenwich Time and the Longitude* (London: Philip Wilson, 1997), 12.
15. Fredric Jameson, *Postmodernism: Or, the Cultural Logic of Late Capitalism* (Durham, N.C.: Duke University Press, 1993), 211.
16. Quoted in Howse, *Greenwich Time,* 108, 109.
17. By 1858, Greenwich time was traveling as far afield as Glasgow, Calcutta, Sydney, and Quebec, where time balls had been established to register Greenwich time by telegraph signals.
18. Quoted in McClintock, *Imperial Leather,* 34.
19. Ibid., 46, 168.

20. Rubinstein, *Before the Suffragettes,* 15.
21. Quoted in ibid., 25, 26, 16, 18.
22. Harrison, *Separate Spheres,* 147.
23. Emmeline Pankhurst, "The Argument of the Broken Pane," in *Speeches and Trials,* 144.
24. Fawcett, *What I Remember,* 179, 182.
25. See Harrison, *Separate Spheres.*
26. Quoted in McClintock, *Imperial Leather,* 118.
27. Quoted in P. Hollis, "Memorandum Relating to the Occurrence (Explosion) in Greenwich Park," February 15, 1894, Royal Greenwich Observatory Archives, Cambridge.
28. "Explosion in Greenwich Park. A Supposed Anarchist Killed. Alleged Conspiracy in London," *Daily Graphic,* February 16, 1894.
29. "The Bomb Explosion in Greenwich Park," *Kentish Mercury,* February 23, 1894.
30. W. J. Thackeray, "Report on the Explosion in Greenwich Park on the Afternoon of Thurs. February 15, 1894," Royal Greenwich Observatory Archives, Cambridge.
31. "Anarchist Conspirators in London," *Illustrated London News,* February 24, 1894.
32. Ibid.
33. "Anarchist Attempt to Blow up Greenwich Observatory: Explosion of a Bomb," *Morning,* February 16, 1894.
34. "Anarchists in London," *Evening News and Post,* February 16, 1894; "Bombs and Anarchy," *Evening News and Post,* February 16, 1894.
35. "Bombs and Anarchy," February 16, 1894.
36. "Are Anarchists Lunatics?" *Evening News and Post,* February 16, 1894.
37. Joseph Conrad, *The Secret Agent* (London: Penguin, 1990), 39. (Subsequent quotations of Conrad are from this novel and appear in the text as page numbers in parentheses.)
38. McClintock, *Imperial Leather,* 120.
39. Christopher GoGwilt, *The Invention of the West: Joseph Conrad and the Double-Mapping of Europe and Empire* (Stanford: Stanford University Press, 1995), 179.
40. Quoted in Lloyd, *Suffragettes International,* 87. Specifically, Lombroso suggested that criminals carried "physical evidence" of features inherited from more "primitive" stages of human evolution, such as abnormal dimensions of the skull and jaw. See Nancy A. Harrowitz, *Antisemitism, Misogyny, and the Logic of Cultural Difference: Cesare Lombroso and Matilde Serao* (Lincoln: University of Nebraska Press, 1994).
41. "Startling Explosion at Greenwich," *Echo,* February 16, 1894.
42. Emmeline Pankhurst, "Kill Me, or Give Me My Freedom!" in *Speeches and Trials,* 316.
43. Emmeline Pankhurst, "Address at Hartford," in *Speeches and Trials,* 324.
44. Emmeline Pankhurst, "The Women's Insurrection," in *Speeches and Trials,* 290.
45. Emmeline Pankhurst, "Great Meeting in the Albert Hall," in *Speeches and Trials,* 278.
46. Harrison, *Separate Spheres,* 78.
47. F. W. Dyson, Astronomer Royal, Secretary of the Admiralty, letter, February 24, 1913, Royal Greenwich Observatory Archives, Cambridge.
48. Quoted in Harrison, *Separate Spheres,* 189.

49. Frantz Fanon, *The Wretched of the Earth* (New York: Grove, 1986), 77.
50. Elizabeth Grosz, *Space, Time, and Perversion* (New York: Routledge, 1995), 135.

CHAPTER 2   PUNDIT A AND THE TRANS-HIMALAYAN SURVEYS

1. Thomas Holdich, "The Use of Practical Geography Illustrated by Recent Frontier Operations," *Geographical Journal* 13, 5 (May 1899): 466, 473, 467. Triangulation, which is based upon the coordinates of latitude and longitude, enabled surveyors to quickly and accurately determine positions by measuring one side of a triangle (called a "base line"), then establishing the location of the two other sides through the angle between this line and a fixed point on the horizon. Because triangulation participated in the seemingly "incontrovertible" discourse of latitude and longitude, it was believed to be invested with the language of impartiality.
2. Clements R. Markham, *A Memoir on the Indian Surveys* (London: Allen, 1878), 59–60.
3. Showell Styles, *The Forbidden Frontiers: The Survey of India from 1765 to 1949* (London: Hamish Hamilton, 1970), 60.
4. Norman J. W. Thrower, *Maps and Civilization: Cartography in Culture and Society* (Chicago: University of Chicago Press, 1996), 60.
5. Quoted in Thomas G. Montgomerie, "On the Geographical Position of Yarkund and Other Places in Central Asia," *Proceedings of the Royal Geograhical Society* 10 (1866): 164.
6. Styles, *Forbidden Frontiers*, 60.
7. L. J. L. Dundas, *India: A Bird's Eye View* (London: Constable, 1924), 63.
8. J. Didion, quoted in David Spurr, *The Rhetoric of Empire: Colonial Discourse in Journalism, Travel Writing, and Imperial Administration* (Durham, N.C.: Duke University Press, 1993), 101.
9. Styles, *Forbidden Frontiers*, 109.
10. Spurr, *Rhetoric of Empire*, 32.
11. Thomas G. Montgomerie, "Report of the Trans-Himalayan Explorations during 1867," *Proceedings of the Royal Geographical Society* 13, 3: 196–198.
12. Markham, *Memoir on the Indian Surveys*, 148.
13. Ibid., 149. According to Thomas G. Montgomerie, the word pundit "simply meant one who had read the 'shasters' or sacred books on the Hindoos. A pundit was simply then an educated Hindoo." See "Report on the Trans-Himalayan Explorations, in Connexion with the Great Trigonometrical Survey of India, during 1865–7," *Proceedings of the Royal Georgraphical Society* 12, 3 (July 15, 1868): 169
14. Markham, *Memoir on the Indian Surveys*, 155, 157.
15. Styles, *Forbidden Frontiers*, 118, 123.
16. Markham, *Memoir on the Indian Surveys*, 150.
17. Montgomerie, "Report on the Trans-Himalayan Explorations, in Connexion with the Great Trigonometrical Survey of India, during 1865–7: Route Survey made by Pundit —, from Nepal to Lhasa, and Thence Through the Upper Valley of the Brahmaputra to Its Source," *Proceedings of the Royal Geographical Society Journal* 12, 3 (July 15, 1868): 155. Subsequent citations of this work in this chapter appear in the text as page numbers in parentheses.
18. Montgomerie, "Report of the Trans-Himalayan Explorations during 1867," 196.

19. Markham, *Memoir on the Indian Surveys*, 161.
20. Mark Paffard, *Kipling's Indian Fiction* (New York: St. Martin's, 1989), 83.
21. Rudyard Kipling, *Kim* (New York: Penguin, 1984), 8. Subsequent quotations from *Kim* in this chapter are identified in the text by page number in parentheses.
22. Abdul JanMohamed, "The Economy of Manichean Allegory: The Function of Racial Difference in Colonialist Literature," *Critical Inquiry* 12 (autumn 1985): 78.
23. Patrick Williams, "Kim and Orientalism," in *Colonial Discourse and Post-Colonial Theory: A Reader*, ed. Patrick Williams and Laura Chrisman (New York: Columbia University Press, 1994), 488.
24. E. Kay Robinson, "Kipling in India," in *Kipling: Interviews and Recollections*, ed. Harold Ore (Totowa, N. J.: Barnes and Noble, 1983), 1:73, 72.
25. Markham, *Memoir on the Indian Surveys*, 167.
26. Montgomerie, "On the Geographical Position of Yarkund," 187.
27. Dundas, *India*, 44.
28. Holdich, "Use of Practical Geography," 473.
29. Paffard, *Kipling's Indian Fiction*, 80.
30. Williams, "Kim and Orientalism," 481.
31. Homi Bhabha, *The Location of Culture* (New York: Routledge, 1994), 81–82.
32. For details of Younghusband's Tibet expedition of 1903–1904, see George Seaver's *Francis Younghusband: Explorer and Mystic* (London: Murray, 1952).
33. "Government Policy on Tibet," statement from the Foreign and Commonwealth Office, January 1994, *http://www.earthlight.co.nz/users/sonamt/Tibet/TibetFacts11.html>* .
34. Montgomerie, "Report on the Trans-Himalayan Explorations," 165.

CHAPTER 3   THE GENDERED AERIAL PERSPECTIVE

1. John Noble Wilford, *The Mapmakers: The Story of Great Pioneers in Cartography from Antiquity to the Space Age* (New York: Vintage, 1982), 233.
2. Though the French were the first to mass-produce aerial cameras, Theodore Scheimpflug of Austria is credited with developing the first aerial mapping camera in 1906. See George W. Goddard, *Overview: A Life-Long Adventure in Aerial Photography* (Garden City, N.Y.: Doubleday, 1969), 9.
3. Clarence Winchester and F. L. Wills, *Aerial Photography: A Comprehensive Survey of Its Practice and Development* (London: Chapman and Hall, 1928); Norman J. W. Thrower, *Maps and Civilization: Cartography in Culture and Society* (Chicago: University of Chicago Press, 1996), 163.
4. Aerofilms, *The Aerofilms Book of Aerial Photographs* (London: Aerofilms, 1965), 314. Wills was an observer with the Royal Naval Air Service in World War I when he conceived the idea of this commercial enterprise, which would go largely unrivaled until 1934. During World War II, Aerofilms formed the basis of the Allied Photo Interpretation Unit (315). In 1921, Aerofilms began mapping London for road-building projects, as well as traffic congestion and subway development.
5. Winchester and Wills, *Aerial Photography*, ix.
6. Susan Ware, *Still Missing: Amelia Earhart and the Search for Modern Feminism* (New York: Norton, 1993), 62.
7. Virginia Woolf, *Mrs. Dalloway* (New York: Harcourt Brace Jovanovich, 1925), 41, 5, 232.

8. Ibid., 29, 40.
9. Virginia Woolf, "Kew Gardens," in *The Virginia Woolf Reader,* ed. Mitchell A. Leaska (San Diego: Harcourt Brace Jovanovich, 1984), 167.
10. Goddard, *Overview,* 9, 8.
11. Winchester and Wills, *Aerial Photography,* 8.
12. Lowell Thomas, *European Skyways: The Story of a Tour of Europe by Airplane* (Boston: Houghton Mifflin, 1927), 26.
13. Paul Virilio, *War and Cinema: The Logistics of Perception,* trans. Patrick Camiller (New York: Verso, 1989), 1, 4.
14. Thomas, *European Skyways,* 32–33, 107.
15. Ibid., 147, 2.
16. Dean Jaros, *Heroes Without Legacy: American Airwomen, 1912–1944* (Niwot: University Press of Colorado, 1993), 15, 13.
17. Amelia Earhart, *The Fun of It: Random Records of My Own Flying and of Women in Aviation* (New York: Harcourt Brace, 1932), 201.
18. Jaros, *Heroes Without Legacy,* 14.
19. Earhart, *The Fun of It,* 152.
20. Ibid., 26.
21. "Aviatrix Must Sign Away Life to Learn Trade," *Chicago Defender,* October 8, 1921, 2.
22. Quoted in Elizabeth Hadley Freydberg, *Bessie Coleman: The Brownskin Lady Bird* (New York: Garland, 1994), 93.
23. Quoted in Doris L. Rich, *Queen Bess: Daredevil Aviator* (Washington, D.C.: Smithsonian Institution Press, 1993), 103.
24. Virginia Woolf, *The Voyage Out,* ed. Elizabeth Heine (London: Hogarth, 1990), 136.
25. Woolf, *Mrs. Dalloway,* 40.
26. Jaros, *Heroes Without Legacy,* 28; Earhart, quoted in ibid., 43.
27. Quoted in Ware, *Still Missing,* 16–18.
28. Ibid., 43, 168.
29. Earhart, *The Fun of It,* 106.
30. Quoted in Ware, *Still Missing,* 63.
31. Quoted in Jaros, *Heroes Without Legacy,* 136, 73.
32. Anne Morrow Lindbergh, *Bring Me a Unicorn: Diaries and Letters of Anne Morrow Lindbergh, 1922–1928* (New York: Harcourt Brace Jovanovich, 1971), 147.
33. Quoted in Jaros, *Heroes Without Legacy,* 72. The page numbers in parentheses in the paragraphs that follow refer to this work.
34. Quoted in Shirley Render, *No Place for a Lady: The Story of Canadian Women Pilots, 1928–1992* (Winnipeg: Portage and Main, 1992), 23, 46, 43, 124.
35. Jaros, *Heroes Without Legacy,* 192–193.
36. Earhart, *The Fun of It,* 173.
37. Freydberg, *Bessie Coleman,* 84, 106.
38. I. N. Phelps Stokes and Daniel C. Haskell, *American Historical Prints: Early Views of American Cities, etc.* From the Phelps Stokes and Other Collections (New York: New York Public Library, 1932), xv; Bourke-White
39. quoted in ibid., 189.
40. Anne Morrow Lindbergh, *Hour of Gold, Hour of Lead: Diaries and Letters of Anne Morrow Lindberg, 1929–1932* (New York: Harcourt Brace Jovanovich, 1973), 93.
41. Sara E. Johnson, *To Spread Their Wings: On the Fiftieth Anniversary of the R.C.A.F. (1941–1991)* (Spruce Grove, Alberta: Saraband Productions, 1990), 58, 60–61.
42. Ibid., 31, 59.

43. Ibid., 110, 118, 53.
44. Lindbergh, *Bring Me a Unicorn,* 216, 246.
45. Thomas, *European Skyways,* 57.
46. Beryl Markham, *West with the Night* (Boston: Houghton Mifflin, 1942), 10.
47. Thomas, *European Skyways,* 4.
48. Earhart, *The Fun of It,* 46.
49. Quoted in Virilio, *War and Cinema,* 74.
50. Markham, *West with the Night,* 176, 153, 152, 185.
51. Lindbergh, *Bring Me a Unicorn,* 106–107, 233, 161, 159.
52. Graham Swift, *Out of This World* (London: Picador, 1997), 189.
53. Walter Benjamin, *Illuminations,* trans. Harry Zohn (New York: Schocken, 1969), 242.
54. Melville C. Branch, *City Planning and Aerial Information,* Harvard City Planning Studies, vol. 17 (Cambridge: Harvard University Press, 1971), 35.
55. Ibid., 99, 38, 13.
56. David Harvey, *The Condition of Postmodernity: An Enquiry into the Origins of Cultural Change* (Cambridge: Blackwell, 1990), 27.
57. Richard Muir, *History from the Air* (London: Michael Joseph, 1983), 195.
58. For further information on the early links between aerial photography and city planning, see M. Pick, "Aerial Photography," 71–86, and J. E. Brittenden, "Surveying for Town Planning by the Use of Aerial Photographs," 86–87, both in *Garden Cities and Town Planning Magazine* 10, 4 (April 1920). See also G. W. Hayler, "The Airplane and City Planning," *American City* 23, 6 (December 1920): 575–579, and Nelson P. Lewis, "A New Aid in City Planning," *American City* 26, 3 (September 1922): 253–255.
59. Tim Bayliss-Smith and Susan Owens, eds., *Britain's Changing Environment from the Air* (Cambridge: Cambridge University Press, 1990), 7.
60. Swift, *Out of This World,* 189.
61. Thomas, *European Skyways,* 238.
62. Quoted in Bayliss-Smith and Owens, *Britain's Changing Environment,* 12.
63. See Goddard, *Overview,* 101.
64. Branch, *City Planning and Aerial Information,* 94.
65. Quoted in Melville C. Branch, *Aerial Photography in Urban Planning and Research,* Harvard City Planning Studies, vol. 14 (Cambridge: Harvard University Press, 1948), 7, 12.
66. Goddard, *Overview,* 60, 63, 109, 63, 107.
67. Ibid., 149, 130, 219.
68. Patrick H. McHaffie, "Manufacturing Metaphors: Public Cartography, the Market, and Democracy," in *Ground Truth: The Social Implications of Geographic Information Systems,* ed. John Pickles (New York: Guilford, 1995), 121.
69. Fredric Jameson, *Postmodernism: Or, the Cultural Logic of Late Capitalism* (Durham, N.C.: Duke University Press, 1993), 51.
70. Benjamin, *Illuminations,* 222–223.
71. Woolf, *Mrs. Dalloway,* 131.
72. Sigmund Freud, "The Uncanny," trans. Alix Strachey, in *Collected Papers: Papers on Metapsychology: Papers on Applied Psycho-Analysis* (London: Hogarth, 1934), 4:387
73. Branch, *Aerial Photography,* 10.
74. Benjamin, *Illuminations,* 223.
75. Aerofilms, *Aerofilms Book of Aerial Photographs,* 316.

76. Gray et al., "Cyborgology: Constructing the Knowledge of Cybernetic Organisms," in *The Cyborg Handbook,* ed. Chris Hables Gray (New York: Routledge, 1995), 4.
77. Markham, *West with the Night,* 283.

CHAPTER 4   AIR CONTROL, ZERZURA, AND THE
              MAPPING OF THE LIBYAN DESERT

1. Claudio G. Segré, *Italo Balbo: A Fascist Life* (Berkeley: University of California Press, 1987), 193.
2. Marc K. Dippold, "Air Occupation: Asking the Right Questions," *Airpower Journal* 11, 4 (winter 1997): 69–84, 78.
3. David W. Parsons, "British Air Control: A Model for the Application of Air Power in Low-Intensity Conflict?" *Airpower Journal* 8, 2 (summer 1994): 27–39, 31.
4. Kenneth J. Alnwick, "Perspectives on Air Power at the Low End of the Conflict Spectrum," *Air University Review* 35, 3 (March–April 1984): 17–28.
5. Desmond Hansen, "The Enigma of Lawrence," *Journal of Historical Review* 2, 3: 283–287.
6. Segré, *Italo Balbo,* 154.
7. See Lisa Anderson, "Libya and American Foreign Policy," in *Middle East Journal* 36 (autumn 1982), 517–518.
8. Quoted in "The Lion of the Desert: A Tribute to Omar Mokhtar," <*http://cyberlabs.com.my/lion/*>.
9. K. Mattawa, *Isma'ilia Eclipse* (New York: Sheep Meadow, 1997), 35–36.
10. Richard A. Bermann, "Historic Problems of the Libyan Desert: A Paper Read with the Preceding [G. J. Penderel, "The Gilf Kebir"] at the Evening Meeting of the Society on 8 January 1934," *Geographical Journal* 83, 6 (1934): 457.
11. Segré, *Italo Balboa,* 295.
12. R. A. Bagnold, *Libyan Sands: Travel in a Dead World* (London: Hodder and Stoughton, 1935), 293.
13. Bermann, "Historic Problems," 457.
14. G. J. Penderel, "The Gilf Kebir: A Paper Read with the Following [Richard A. Bermann, "Historic Problems"] at the Evening Meeting of the Society on 8 January 1934," *Geographical Journal* 83, 6 (1934): 450.
15. Dorothy Clayton-East-Clayton, "The Lost Oasis: Across the Sand Sea," *London Times,* September 16, 1933, 11.
16. Bagnold, *Libyan Sands,* 242–243.
17. Ibid., 255; Bermann, "Historic Problems," 464.
18. O. Wingate, "In Search of Zerzura," *Geographical Journal* 83, 4 (1934): 281. According to Bagnold, the "last big camel journey to be made by Europeans" was in 1927 (*Libyan Sands,* 240).
19. Quoted in Segré, *Italo Balboa,* 309.
20. Tyler McLeod, "Patient Author Impressed by Movie," *New York Express,* November 3, 1996.
21. Michael Ondaatje, *The English Patient* (New York: Vintage, 1993), 261. Subsequent citations of this work in this chapter appear in the text as page numbers in parentheses.
22. It isn't until "1933 or 1934" that Almásy begins to use "A-type Ford cars with box bodies" on his expeditions (139).

23. Jean Howard, letter to the Royal Geographical Society, 1985, "Kufara Expedition, 1933—Italian Official Documents Referring to Clayton, P. A. and Penderel, H.W.G.J., etc. and the Expeditions to Gilf Kebir, Part 1," Royal Geographical Society Archives. Howard knew Almásy between 1941 and 1942.

24. Willem Dafoe, "Michael Ondaatje on *The English Patient,*" *Bomb,* winter 1997, 19.

25. "A Memoir on the Geography of the North-Eastern part of Asia, and on the Question Whether Asia and America Are Contiguous, or Are Separated by the Sea," *Quarterly Review* 18 (May 1818): 434–435.

26. Christopher L. Miller, *Blank Darkness: Africanist Discourse in French* (Chicago: University of Chicago Press, 1985), 6; Herodotus quoted in ibid., 3.

27. Herodotus, *The Histories,* trans. A. D. Godley, Loeb Classical Library, 387.

28. T. E. Lawrence, *Seven Pillars of Wisdom: A Triumph* (Harmondsworth, Eng.: Penguin, 1962), 28.

29. Bermann, "Historic Problems," 456.

30. Penderel, "The Gilf Kebir," 457.

31. Bagnold, *Libyan Sands,* 345.

32. Bermann, "Historic Problems," 459.

33. Ondaatje's scene is actually repeated nearly word-for-word from Bermann's account in the *Geographical Journal:* "I shall always remember a conference we had in the old Senussi fortress of Djof. . . . Almásy was asking questions of Ibrahim, and old Tebu who was by profession a caravan guide: a snake-like, mysterious old man, quite black. . . . The influence of the Senussi creed is still strong in Kufara, and not to reveal the secrets of the desert to the stranger had been one of the foremost doctrines of the Senussiya" (Bermann, "Historic Problems," 459).

34. Winona LaDuke, "We Are Still Here: The Five Hundred Years Celebration," *Sojourners,* October 1991, 12.

35. David Spurr, *The Rhetoric of Empire: Colonial Discourse in Journalism, Travel Writing, and Imperial Administration* (Durham, N.C.: Duke University Press, 1993), 107.

36. Spurr, *Rhetoric of Empire,* 130.

37. Lawrence, *Seven Pillars of Wisdom,* 38.

38. Bagnold, *Libyan Sands,* 283.

39. Ibid.

40. L. Almásy, *The Unknown Sahara,* quoted in Zsolt Török, "Desert Love: Lászlo Almásy, the Real English Patient," *Mercator's World* 2, 5 (September–October 1997): 43.

41. J. Baudrillard, *America* (New York: Verso, 1988), 63. In an interview entitled "Vivisecting the 90s," Baudrillard again delineates his notion of deserts: "Deserts are a metaphor for disappearing objects, evanescence beyond culture." Trans. C. Bayard and G. Knight, *CTHEORY,* March 8, 1995 <*http:// www.ctheory.com*>. For further discussion on Foucault and deserts, see W. E. Paden, "Theaters of Humility and Suspicion: Desert Saints and New England Puritans," in *Technologies of the Self: A Seminar with Michel Foucault,* ed. L. H. Martin, H. Gutman, and P. H. Hutton (London: Tavistock, 1988), 64–79. Also, Gilles Deleuze and Félix Guattari equate the "schizophrenic voyage" with "the meaning of American frontiers: something to go beyond, limits to cross over, flows to set in motion, noncoded spaces to enter" (*Anti-Oedipus: Capi-*

talism and Schizophrenia, trans. R. Hurley, M. Seem, and H. R. Lane [Minneapolis: University of Minnesota Press, 1983, 224).

42. Deleuze and Guattari, *Anti-Oedipus,* 194.
43. Bagnold, *Libyan Sands,* 327; Bermann, "Historic Problems," 463.
44. Bagnold, *Libyan Sands,* 347, 283.
45. Quoted in Spurr, *Rhetoric of Empire,* 146.
46. Miller, *Blank Darkness,* 19.
47. R. W. Bagnold, letter to G. W. Murray, June 8, 1932, P. A. Clayton file, Royal Geographical Society Archives, London.
48. G. J. Penderel explains: "Almásy with one of the natives had . . . made a reconnaissance on foot into the Gilf and had found the third wadi. The credit of the discovery must rest wholly with Almásy" ("The Gilf Kebir," 455).
49. Homi Bhabha, *The Location of Culture* (New York: Routledge, 1994), 25.
50. Bagnold, *Libyan Sands,* 347.
51. Jean Baudrillard, "Game with Vestiges," *On the Beach* 5 (winter 1984): 24.
52. Penderel, "The Gilf Kebir," 450.
53. She wrote, "After my husband's death I determined to finish the work of discovery we had begun." Clayton-East-Clayton, "The Lost Oasis."
54. "Obituary: Lady Clayton East Clayton," *London Times,* September 16, 1933.
55. This letter from Francis Rodd to Ahmed Pasha Hassanein, dated January 25, 1933, is in RGS (letters) file no. 2, January 1, 1933–December 31, 1934, Royal Geographical Society Archives, London.
56. This letter to Francis Rodd from Ahmed Pasha Hassanein, dated January 14, 1933, is in RGS (letters) file no. 2, January 1, 1933–December 31, 1934, Royal Geographical Society Archives.
57. D. Clayton-East-Clayton, "The Last Oasis," *Times,* September 16, 1933; G. J. Penderel wrote: "It was here [Kufara] we became aware that Clayton had received most able reinforcements in the shape of Lady Clayton-East-Clayton and Commander Roundell's expedition. . . . They left word for us that they had discovered a second wadi" ("The Gilf Kebir," 454).
58. "Lady Clayton Killed. An Aeroplane Accident," *Times,* September 16, 1933; "The Brooklands Accidents. Inquests on Three Victims. Death by Misadventure," *Times,* September 22, 1933.
59. "Two Fatal Days at Brooklands. Graphic Inquest Stories. Lady Clayton's Leap to Death, Racing Motorist Who Crashed at 110 Miles an Hour," *Daily Express,* September 19, 1933.
60. "Tragic Death of Vicar's Daughter Killed at Brooklands. Funeral at Leversotck Green," *Hemel Hempstead Gazette,* September 23, 1933.
61. Ibid.
62. "Rare Germ Kills Young Baronet: Delayed Action Peril of the Desert," *Daily Express,* September 3, 1932.
63. William Hickey, "Killed Yesterday in Unusual Airplane Crash: Lady Clayton East Clayton," *Daily Express,* September 16, 1933.
64. On April 26, 1934, a letter from F. Rodd to A. Hinks stated: "Lady Clayton's photos of silica glass lying in the sand have been lost." RGS (letters) file no. 2, January 1, 1933–December 31, 1984, Royal Geographical Society Archives.
65. Ibid. The signature on this letter, dated March 6, 1933, is illegible.
66. Bagnold, *Libyan Sands,* 333.
67. See Raoul Schrott and Michael Farin, preface to *Schwimmer in der Wüste,* by Lázló Almásy (Innsbruck: Haymon, 1997), 16.

68. G. W. Murray, "Obituary: Ladislas Almásy," *Geographical Journal* 127, 2 (1951): 254.

69. Steven Totosy de Zepetnek, "Michael Ondaatje's *The English Patient*, 'History,' and the Other," *CLCWeb: Comparative Literature and Culture* 1, 4 (1999) <*http://www.arts.ualberta.ca/clcwebjournal/clcweb99–4/contents99–4.html*>.

70. Samir Raafat, "*The English Patient*: Egypt's Celebrated Hungarian Brothers," *Egyptian Mail*, June 7, 1997.

71. Quoted in Zepetnek, "Michael Ondaatje's *The English Patient.*"

72. Schrott and Farin, preface, 8.

73. "Disappointing Original," *Indian Express*, August 25, 1997.

74. Quoted in Török, "Desert Love," 47.

75. Quoted in Donovan Webster, "Journey to the Heart of the Sahara," *National Geographic*, March 1999, 24.

76. Quoted in Majid Khadduri, *Modern Libya: A Study in Political Development* (Baltimore: Johns Hopkins University Press, 1963), 30, 35.

77. Penelope Lively, *Moon Tiger* (New York: HarperPerennial, 1987), 116.

78. Winchester and Wills, *Aerial Photography*, 7.

79. Michael Ondaatje, *Running in the Family* (New York: Vintage, 1993), 188, 53, 172.

80. Prakrti, "The Breach: Three Sri Lankan-Born Writers at the Crossroads," *Lanka Outlook*, summer 1997.

81. Aijaz Ahmad, "Orientalism and After," in *Colonial Discourse and Post-Colonial Theory*, ed. Patrick Williams and Laura Chrisman (New York: Columbia University Press, 1994), 169.

82. Bhabha, *Location of Culture*, 193, 33.

83. Hsen Larbi, "The Amazigh World Congress," *Amazigh Voice* 4–5, 3–1, December 1995–March 1996 <*http://www.umd.edu/~sellami/DEC95/congress.html*>.

84. Ahmad, "Orientalism and After," 170.

85. Quoted in Dafoe, "Michael Ondaatje," 19.

86. Ondaatje, *Running in the Family*, 41.

87. Gary Kamiya, "An Interview with Michael Ondaatje," *Salon*, November 1996 <*http://www.salon.com/nov96/ondaatje961118.html*>.

88. Dippold, "Air Occupation," 78.

CHAPTER 5   CANADIAN CARTOGRAPHY, POSTNATIONALISM, AND THE GREY OWL SYNDROME

1. George W. Goddard, *Overview: A Life-Long Adventure in Aerial Photography* (Garden City, N.Y.: Doubleday, 1969), 382.

2. The term "paperclip" referred to a special paperclip that was attached to the files of these scientists to indicate that they had been purged of Nazi war crimes. For a detailed history of Project Paperclip, see Tom Bower, *The Paperclip Conspiracy* (Boston: Little, Brown, 1987), or Linda Hunt, *Secret Agenda: The United States Government, Nazi Scientists, and Project Paperclip, 1945 to 1990* (New York: St. Martin's, 1991). Also, see Eugene M. Emme, *Aeronautics and Astronautics: An American Chronology of Science and Technology in the Exploration of Space, 1915–1960* (Washington, D.C.: NASA, 1961).

3. Nicholas M. Short, Sr. "Remote Sensing and Image Interpretation and Analysis," Code 935, Goddard Space Flight Center, NASA.

4. L. W. Morley, "Remote Sensing Then and Now," Canada Center for Remote Sensing, <*http://www.ccrs.nrcan.gc.ca/ccrs/org/history/historye.html*>.
5. Margaret Atwood and Victor-Lévy Beaulieu, *Two Solicitudes: Conversations,* trans. Phyllis Aronoff and Howard Scott (Toronto: McClelland and Stewart, 1998), 117.
6. G. Ball, *The Discipline of Power* (Boston: Atlantic/Little, Brown, 1968), 113.
7. Atwood and Beaulieu, *Two Solicitudes,* 104.
8. Margaret Atwood, *Survival: A Thematic Guide to Canadian Literature* (Toronto: Anansi, 1972), 35.
9. Bart Moore-Gilbert, *Postcolonial Theory: Contexts, Practices, Politics* (New York: Verso, 1997), 10.
10. Atwood and Beaulieu, *Two Solicitudes,* 103.
11. Leela Gandhi, *Postcolonial Theory: A Critical Introduction* (New York: Columbia University Press, 1998), 124.
12. "GIS World Interview—Roger Tomlinson: The Father of GIS," *GIS World* 9, 4 (April 1996): 56–60.
13. In 1919, the Air Board of Canada was formed for the purpose of aerial photography, which provided a type of quick resource survey of inaccessible areas and was considered "most useful in the mapping of inland water country, vast swamp districts and impenetrable forest regions." Clarence Winchester and F. L. Wills, *Aerial Photography: A Comprehensive Survey of Its Practice and Development* (London: Chapman and Hall, 1928), ix.
14. In 1970, the first GIS conference in the world took place in Ottawa, funded by UNESCO. "*GIS World* Interview."
15. "Land Capability Rating," from the Canada Land Inventory web page, <*http://geogratis.cgdi.gc.ca/cli*>.
16. "Overview of Classification Methodology for Determining Land," ibid.
17. Margaret Atwood, *Surfacing* (New York: Doubleday, 1998), 150. Subsequent citations of this work in this chapter appear in the text as page numbers in parentheses.
18. Margaret Atwood, "Approximate Homes," in *Writing Home: A PEN Canada Anthology,* ed. Constance Rooke (Toronto: McClelland and Stewart, 1997), 6.
19. Margaret Atwood, *Strange Things: The Malevolent North in Canadian Literature* (Oxford: Clarendon, 1995), 2, 19.
20. Mary Lannon, "Margaret Atwood," *Writers Online* 3, 1 (fall 1998), <*http://www.albany.edu/writers-inst/online.html*>.s
21. See David Ward, "Surfacing: Separation, Transition, Incorporation," in *Margaret Atwood: Writing and Subjectivity: New Critical Essays,* ed. Colin Nicholson (New York: St. Martin's, 1994), 94–118.
22. Margaret Atwood, *The Journals of Susanna Moodie* (New York: Houghton Mifflin, 1997); Atwood and Shields are quoted in the foreword by David Staines, x.
23. Susanna Moodie, *Roughing It in the Bush: Life in the Backwoods* (New York: Lovell, 1884), 27, 46.
24. Ibid., 28.
25. Barry Lopez, *Arctic Dreams: Imagination and Desire in a Northern Landscape* (New York: Scribner's, 1986), 287.
26. Atwood, *Strange Things,* 59–60.
27. Ibid., 2.

28. Ibid., 60.
29. Atwood and Beaulieu, *Two Solicitudes,* 11.
30. Darlene Wroe, "Political Leaders and Industry Lead Rally to Tell Opponents to Go Home," *Land Caution News,* October 2, 1996.
31. Atwood, *Strange Things,* 91, 4, 108. She writes that the north is "a sort of icy and savage femme fatale who will drive you crazy and claim you for her own. But what happens if the 'you' that is being driven and/or claimed is not a man, but a woman?"
32. Marian Engel, *Bear* (Toronto: New Canadian Library, 1991), 49, 137.
33. Max Horkheimer and Theodor Adorno, *Dialectic of Enlightenment,* trans. John Cumming (New York: Seabury, 1972), 230.
34. Frederick Jackson Turner, *The Frontier in American History* (New York: Holt, Rinehart and Winston, 1962), 4.
35. Richard Jonasse, "Landscapes of Facts: GIS and the Topography of Power," presented at the International Conference on Geographic Information and Society," June 20–22, 1999, University of Minnesota, Minneapolis; Jean Baudrillard, *Simulations,* trans. Paul Foss, Paul Patton, and Philip Beitchman (New York: Semiotext(e), 1983), 149.
36. Wayne Madsen, "Protecting Indigenous Peoples' Privacy from 'Eyes in the Sky,'" *Proceedings of the Conference on Law and Information Policy for Spatial Databases,* ed. H. J. Onsrud (Orono, Maine: National Center for Georgraphic Information and Analysis, 1994), 223–231.
37. Peter J. Taylor and Ronald J. Johnston, "Geographic Information Systems and Geography," in *Ground Truth: The Social Implications of Geographic Information Systems,* ed. John Pickles (New York: Guilford, 1995), 58.
38. Ian Darragh, "Quebec's Quandary," *National Geographic,* November 1997, 71, 64.
39. See the 1998 Recipients for the National Aboriginal Achievement Award at <*http://www.naaf.ca/rec98.html*>.
40. "Six Nations Adapts Traditional Beliefs to New Technology with GIS," *ArcNews* 22, 1 (spring 2000): 11.
41. Frank Duerden and Valerie A. Johnson, "GIS and the Visualization of First Nations Land Selections," in *GIS '93 Symposium Proceedings* (Vancouver, B.C.: Polaris Conferences, 1993), 729.
42. P. Poole, "Indigenous Lands and Power Mapping in the Americas: Merging Technologies," *Native Americas* 15, 4 (winter 1998): 31–43.
43. "Nunavut Planning Commission," <*http://npc.nunavut.ca/eng/npc/*>.
44. Lopez, *Arctic Dreams,* 287.
45. Ibid., 297; McClure quoted in ibid., 287.
46. Roman Frank and David Carruthers, "Networking the Aboriginal Mapping Community in British Columbia," Aboriginal Mapping Network, <*http://www.nativemaps.org/ website*>.
47. Gandhi, *Postcolonial Theory,* 109.
48. Poole, "Indigenous Lands."
49. Atwood, *Strange Things,* 58.
50. David Spurr, *The Rhetoric of Empire: Colonial Discourse in Journalism, Travel Writing, and Imperial Administration* (Durham, N.C.: Duke University Press, 1993), 92–93, 82–83.
51. Atwood, *Strange Things,* 6.

CHAPTER 6   POSTCOLONIAL OCCUPATIONS OF CYBERSPACE

1. Philip K. Dick, *VALIS* (New York: Bantam, 1981), 33.
2. "Philip K. Dick: Exegesis Online" <*http://www.aol-i.com/eoe/exegesis.asp*>.
3. R. J. Toth, "A Life of Fantasy, a Literature of Fantasy," *Wall Street Journal,* April 27, 1999.
4. Philip K. Dick, "I Hope I Shall Arrive Soon," in *The Cyborg Handbook,* ed. Chris Hables Gray (New York: Routledge, 1995).
5. Erik Davis "Techgnosis, Magic, Memory, and the Angels of Information," in *Flame Wars: The Discourse of Cyberculture,* ed. Mark Dery (Durham, N.C.: Duke University Press, 1994), 52–53.
6. Dick, *VALIS,* 5; David Pringle and Andy Robertson, "Bayley J. Barrington Interview," *Interzone* 35 (May 1990): 18.
7. Lawrence Sutin, preface to *In Pursuit of Valis: Selections from the Exegesis,* by Philip K. Dick, ed. Lawrence Sutin (Grass Valley, Calif.: Underwood, 1991).
8. See James M. Robinson, ed., *The Nag Hammadi Library in English* (San Francisco: Harper and Row, 1977).
9. *St. Irenaeus of Lyons: Against the Heresies,* trans. Dominic J. Unger, vol. 1, book 1 (New York: Paulist Press, 1992), 72, 74.
10. Jean Baudrillard celebrates the "radical semiurgy" of postmodern signs and simulations; Michel Foucault advocates the pleasure of "technologies of the self," which some argue include "on-line role playing" and other Internet discourse; and Gilles Deleuze and Félix Guattari suggest liberating the flows of desire with a revolutionary "schizoanalysis" that performs the same function as the computer.
11. Mark Nunes, "Jean Baudrillard in Cyberspace: Internet, Virtuality, and Postmodernity," *Style* 29, 2 (summer 1995): 326.
12. Davis, "Techgnosis," 39.
13. John Pickles, "Towards an Economy of Electronic Representation and the Virtual Sign," in *Ground Truth: The Social Implications of Geographic Information Systems,* ed. John Pickles (New York: Guilford, 1995), 227.
14. Ibid., 238.
15. Amitav Ghosh, *The Calcutta Chromosome* (New York: Avon, 1995), 7. Subsequent citations of this work in this chapter appear in the text as page numbers in parentheses.
16. Salman Rushdie, "At the Auction of the Ruby Slippers," in *East, West* (New York: Vintage International, 1994), 94, 92, 98.
17. G. Roulet, "The New Utopia: Communication Technologies," *Telos* 87 (1992): 40.
18. Amitav Ghosh, *In an Antique Land* (New York: Vintage, 1994), 342.
19. Gauri Viswanathan, "Beyond Orientalism: Syncretism and the Politics of Knowledge," *Stanford Humanities Review* 5, 1 (1995): 32.
20. Arundhati Roy, "The Greater Common Good," in *The Cost of Living* (New York: Modern Library, 1999), 16–17, 19, 23–24, 13, 65.
21. Ibid., 64.
22. "IWMI Is Set to Become the 'Global Water Center,'" <*http://www.cgir.org/iwmi/press/press1.htm*>.
23. James Saynor, "Trapped in a Worldwide Web," *New York Times,* September 14, 1997.
24. Quoted in Davis, "Techgnosis," 48.

25. The structure of *The Calcutta Chromosome* appears to mimic, in many ways, the Heisenburg principle of uncertainty, a law in quantum physics that suggests that the perceiver's position affects the object being perceived—this principle, however, is never explicitly mentioned in the text.

CONCLUSION   POSTMODERN PRIMITIVISM

1. Al Gore, "The Digital Earth: Understanding Our Planet in the Twenty-first Century," remarks at the California Science Center, Los Angeles, January 31, 1998, <*http://www.gik.uni-karlsruhe.de/~zippelt/gis-welt/DigitalEarth.html*>.
2. Ibid.
3. Terry Standley, "GIS Implementation in Developing Countries: A United Nations Perspective," paper presented at the GIS/GPS Conference '97, March 2–4, 1997, Doha, Quatar, <*http://www.gisqatar.org.qa/conf97*>.
4. Arturo Escobar, "Imagining a Post-Development Era? Critical Thought, Development, and Social Movements," *Social Text* 30/31 (1992): 24.
5. Al Gore, "Remarks," presented at the ITU Plenipotentiary Conference in Minneapolis, October 12, 1998, <*http://www.itu.int/newsroom/press/PP98/Documents*>.
6. Ricardo Gómez, Patrik Hunt, and Emmanuelle Lamoureux, "Telecentros en la Mira: Cómo Pueden Contribuir al Desarrollo Social?" [Focus on telecenters: How can they contribute to social development?], *Chasqui: Latin American Review of Communication,* June 1999, 54–58. Translated at: <*http://www.idrc.ca/pan/chasqui.html*>.
7. S. Robinson, "Telecenters in Mexico: Learning the Hard Way," paper presented at "Partnerships and Participation in Telecommunications for Rural Development: Exploring What Works and Why," conference sponsored by the Don Snowden Program, the TeleCommons Development Group, and the Foundation for International Training, University of Guelph, Guelph, Ontario, October 26–27, 1998.
8. Carol Hall, "Gender and GIS," paper presented at the "GIS and Society Workshop," South Haven, Minn., March 2–5, 1996. See also Anne Lloyd and Liz Newell, "Women and Computers," in *Smothered by Invention: Technology in Women's Lives,* ed. Wendy Faulkner and Erik Arnold (London: Pluto, 1985).
9. Peter J. Taylor and Ronald J. Johnston, "Geographic Information Systems and Geography," in *Ground Truth: The Social Implications of Geographic Information Systems,* ed. John Pickles (New York: Guilford, 1995), 59.
10. They write: "The forced removal of blacks from designated 'white' territories and their relocation into urban territories and rural bantustans was planned, organized, and researched through the institutions of the apartheid state." T. M. Harris et al., "Pursuing Social Goals Through Participatory GIS: Redressing South Africa's Historical Political Ecology," in *Ground Truth,* 204.
11. Vandana Shiva, *Staying Alive: Women, Ecology, and Development* (London: Zed, 1989), 85.
12. National Remote Sensing Agency (NRSA), *Guidelines to Use Wasteland Maps* (Hyderabad, India: NRSA, Department of Space, 1991), 2.
13. Graham Dudley, "The Political Ecology of Geographical Information in India," paper presented at the Initiative 19 Specialist Meeting, South Haven, Minn., March 2–5, 1996, <*http://www.geo.wvu.edu/i19/papers/position.html*>.

14. Shiva, *Staying Alive*, 83.
15. Quoted in Escobar, "Imagining a Post-Development Era," 23.
16. Marianna Torgovnick, *Gone Primitive: Savage Intellects, Modern Lives* (Chicago: University of Chicago Press, 1991), 46.
17. V. Vale, *Modern Primitives* (San Francisco: Re/Search, 1989), 4.
18. "Renew the Earthly Paradise," reprinted in *Green Anarchist*, autumn 1995, 6.
19. Fredy Perlman, *Against History, Against Leviathan* (Detroit: Black and Red, 1983).
20. John Moore, "Comin' Home: Defining Anarcho-Primitivism," *Green Anarchist*, summer 1995, 7.
21. Steven Best and Douglas Kellner, *Postmodern Theory* (New York: Guilford, 1991), 116. Baudrillard, however, denies that he is advocating a return to a symbolic order. "We cannot recreate a symbolic order," he writes in *The Disappearance of Art and Politics* (New York: St. Martin's, 1992), 296.
22. Jean Baudrillard, *Symbolic Exchange and Death* (London: Sage, 1993). Lyotard's critique of Baudrillard can be found in *Libidinal Economy*, trans. Iain Hamilton Grant (Bloomington: Indiana University Press, 1993), 106. See also Jean-François Lyotard, *Just Gaming*, trans. Wlad Godzich (Minneapolis: University of Minnesota Press, 1985), and "Lessons in Paganism," trans. David Macey, in *The Lyotard Reader*, ed. Andrew Benjamin (Oxford: Basil Blackwell, 1989).
23. Gilles Deleuze and Félix Guattari, *Anti-Oedipus: Capitalism and Schizophrenia*, trans. R. Hurley, M. Seem, and H. R. Lane (Minneapolis: University of Minnesota Press, 1983), 167, 185–187.
24. Gilles Deleuze and Félix Guattari, *A Thousand Plateaus: Capitalism and Schizophrenia*, trans. Brian Massumi (Minneapolis: University of Minnesota Press, 1987), 557n.56; Edmund Carpenter, "If Wittgenstein Had Been an Eskimo," *Explorations* 12, 3 (1966): 59. Christopher L. Miller criticizes Deleuze and Guattari's primitivist discourse in "The Postidentitarian Predicament in the Footnotes of A Thousand Plateaus: Nomadology, Anthropology, and Authority," *diacritics* 23, 3 (1993): 6–33.
25. Torgovnick, *Gone Primitive*, 185.
26. Edward Said, *Culture and Imperialism* (New York: Random House, 1993), 169–170.
27. P. Poole, "Indigenous Lands and Power Mapping in the Americas: Merging Technologies," *Native Americas* 15, 4 (winter 1998): 41.
28. Frank Duerden and Richard Kuhn, "Indigenous Land-Use Information Project," paper presented at the School of Applied Geography, Ryerson Polytechnic University, Toronto, 1996. This is from the project proposal, available at <*http://www.ryerson.ca/geog/wwwsag/html/indigen.html*>.
29. Poole, "Indigenous Lands," 36.
30. Barry Lopez, *Arctic Dreams: Imagination and Desire in a Northern Landscape* (New York: Scribner's, 1986), 298.
31. C. Olive and D. Carruthers, "Putting TEK into Action: Mapping the Transition," paper presented at "Bridging Traditional Ecological Knowledge and Ecosystem Science," Flagstaff, Ariz., August 13–15, 1998.
32. Tina Cary, "Visions of the Information Industry: Dreams or Nightmares?" paper presented at the GIS/GPS Conference '97, sponsored by the Center for GIS, State of Qatar, Doha, Qatar, March 2–4, 1997, <*http://www.gisqatar.org.qa/conf97*>.
33. Jan-Willem Broekhuysen, preface to *Greenwich Time and the Longitude*, by Derek Howse (London: Philip Wilson, 1997), 9.

Bibliography

ARCHIVAL MATERIALS

Bagnold, R. W. Letter to G. W. Murray, June 8, 1932. P. A. Clayton file. Royal Geographical Society Archives, London.

FWN Astronomer Royal, Secretary of the Admiralty. Letter to the Royal Geographical Society, February 24, 1913. Royal Greenwich Observatory Archives, Cambridge.

Hassanein, Ahmed Pasha. Letter to Francis Ross, January, 14, 1933. RGS (letters) file no. 2, January 1, 1933–December 31, 1934. Royal Geographical Society Archives, London.

Hollis, H. P. "Memorandum Relating to the Occurrence (Explosion) in Greenwich Park," February 15, 1894. Royal Greenwich Observatory Archives, Cambridge.

Howard, Jean. Letter to the Royal Geographical Society, 1985, "Kufara Expedition, 1933—Italian Official Documents Referring to Clayton, P. A. and Penderel, H.W.G.J., etc. and the Expeditions to Gilf Kebir, Part 1." Royal Geographical Society Archives, 1985.

Rodd, Francis. Letter to Ahmed Pasha Hassanein, January 25, 1933. RGS (letters) file no. 2, January 1, 1933–December 31, 1934. Royal Geographical Society Archives, London.

———. Letter to Arthur Hinks, April 26, 1934. RGS (letters) file no. 2, January 1, 1933–December 31, 1934, Royal Geographical Society Archives, London.

Thackeray, W. J. "Report on the Explosion in Greenwich Park on the Afternoon of Thurs. February 15, 1894." Royal Greenwich Observatory Archives, Cambridge.

PRINTED WORKS

Aerofilms. *The Aerofilms Book of Aerial Photographs.* London: Aerofilms, 1965.

Ahmad, Aijaz. "Orientalism and After." In *Colonial Discourse and Post-Colonial Theory,* edited by Patrick Williams and Laura Chrisman. New York: Columbia University Press, 1994.

Alnwick, Kenneth J. "Perspectives on Air Power at the Low End of the Conflict Spectrum." *Air University Review* 35, 3 (March–April 1984): 17–28.

"A Memoir on the Geography of the North-Eastern Part of Asia, and on the Question Whether Asia and America Are Contiguous, or Are Separated by the Sea." *Quarterly Review* 18 (May 1818): 434–435.

Anderson, Benedict. *Imagined Communities: Reflections on the Origin and Spread of Nationalism.* New York: Verso, 1991.

Anderson, Lisa. "Qaddafi's Islam." In *Voices of Resurgent Islam,* edited by John L. Esposito. New York: Oxford University Press, 1983.

Atwood, Margaret. "Approximate Homes." In *Writing Home: A PEN Canada Anthology,* edited by Constance Rooke. Toronto: McClelland and Stewart, 1997.

———. *The Journals of Susanna Moodie.* New York: Houghton Mifflin, 1997.

———. *Strange Things: The Malevolent North in Canadian Literature.* Oxford: Clarendon, 1995

———. *Surfacing.* London: Virago, 1979.

———. *Survival: A Thematic Guide to Canadian Literature.* Toronto: Anansi, 1972.

Atwood, Margaret, and Victor-Lévy Beaulieu. *Two Solicitudes: Conversations.* Translated by Phyllis Aronoff and Howard Scott. Toronto: McClelland and Stewart, 1998.

Ball, G. *The Discipline of Power.* Boston: Atlantic/Little, Brown, 1968.

Bagnold, R. A. *Libyan Sands: Travel in a Dead World.* London: Hodder and Stoughton, 1935.

Baudrillard, Jean. *America.* New York: Verso, 1988.

———. *The Disappearance of Art and Politics.* New York: St. Martin's, 1992.

———. "Game with Vestiges." *On the Beach* 5 (winter 1984): 19–25.

———. *Simulations.* Translated by Paul Foss, Paul Patton, and Philip Beitchman. New York: Semiotext(e), 1983.

———. *Symbolic Exchange and Death.* London: Sage, 1993.

———. "Vivisecting the 90s: An Interview with Jean Baudrillard." Translated by Caroline Bayard and Graham Knight. *CTHEORY,* March 8, 1995, *http://www.ctheory.com,* accessed October 2001.

Bayliss-Smith, Tim, and Susan Owens, eds. *Britain's Changing Environment from the Air.* Cambridge: Cambridge University Press, 1990.

Benjamin, Walter. *Illuminations.* Translated by Harry Zohn. New York: Schocken, 1969.

Bermann, Richard A. "Historic Problems of the Libyan Desert: A Paper Read with the Preceding [G. J. Penderel, "The Gilf Kebir"] at the Evening Meeting of the Society on 8 January 1934." *Geographical Journal* 83, 6 (1934): 456–470.

Best, Steven, and Douglas Kellner. *Postmodern Theory: Critical Interrogations.* New York: Guilford, 1991.

Bhabha, Homi. *The Location of Culture.* New York: Routledge, 1994.

Blais, Marie-Claire. Afterword to *Surfacing,* by Margaret Atwood. Toronto: McClelland and Stewart, 1994.

Borley, S. F. H. *A Review of the Literature on the Use of Geographical Systems in Developing Countries.* Sheffield: University of Sheffield, Department of Information Studies, 1991.

Bower, Tom. *The Paperclip Conspiracy.* Boston: Little, Brown, 1987.

Branch, Melville C. *Aerial Photography in Urban Planning and Research.* Harvard City Planning Studies, vol. 14. Cambridge: Harvard University Press, 1948.

———. *City Planning and Aerial Information.* Harvard City Planning Studies, vol. 17. Cambridge: Harvard University Press, 1971.

Brennan, Timothy. "The National Longing for Form." In *Nation and Narration,* edited by H. Bhabha, 44–70. New York: Routledge, 1990.

Brittenden, J. E. "Surveying for Town Planning by the Use of Aerial Photographs." *Garden Cities and Town Planning Magazine* 10, 4 (April 1920): 86–87.

Carpenter, Edmund. "If Wittgenstein Had Been an Eskimo." *Explorations* 12, 3: 49–64.

Cary, Tina. "Visions of the Information Industry: Dreams or Nightmares?" Paper presented at the GIS/GPS Conference '97. Sponsored by the Center for GIS, State of Qatar, Doha, Quatar, March 2–4, 1997. *http://www.gisqatar.org.qa/conf97,* accessed October 2001.

CCH Macquarie Concise Dictionary of Modern Law. Sydney, Aus.: CCH, 1988.

"Closing Days of the Trial [May 17–22, 1912." In *Speeches and Trials of the Militant Suffragettes: The Women's Social and Political Union,* edited by Cheryl R. Jorgensen-Earp. Madison, N.J.: Fairleigh Dickinson University Press, 1999.

Clynes, Manfred E. "Cyborg II: Sentic Space Travel." In *The Cyborg Handbook,* edited by Chris Hables Gray, 43–54. New York: Routledge, 1995.

Clynes, Manfred E., and Nathan S. Kline. "Cyborgs and Space." In *The Cyborg Handbook,* edited by Chris Hables Gray, 29–34. New York: Routledge, 1995.

Conrad, Joseph. *Heart of Darkness.* New York: Penguin, 1999.

———. *The Secret Agent.* London: Penguin, 1990.

Crommelin, Michael. "Mabo Explained." *Law Institute Journal* 67, 9 (September 1993), 809–811.

Curry, Michael R. "Geographic Information Systems and the Inevitability of Ethical Inconsistency." In *Ground Truth: The Social Implications of Geographic Information Systems,* edited by John Pickles, 68–87. New York: Guilford, 1995.

Dafoe, Willem. "Michael Ondaatje on *The English Patient.*" *Bomb,* winter 1997, 14–19.

Darragh, Ian. "Quebec's Quandary." *National Geographic,* November 1997, 46–67!

Davis, Erik. "Techgnosis, Magic, Memory, and the Angels of Information." In *Flame Wars: The Discourse of Cyberculture,* edited by Mark Dery, 29–60. Durham, N.C.: Duke University Press, 1994.

Defoe, Daniel. *Robinson Crusoe.* New York: Bantam, 1981.

Deleuze, Gilles, and Félix Guattari. *Anti-Oedipus: Capitalism and Schizophrenia.* Translated by R. Hurley, M. Seem, and H. R. Lane. Minneapolis: University of Minnesota Press, 1983.

———. *A Thousand Plateaus: Capitalism and Schizophrenia.* Translated by Brian Massumi. Minneapolis: University of Minnesota Press, 1987.

Dick, Philip K. "I Hope I Shall Arrive Soon." In *The Cyborg Handbook,* edited by Chris Hables Gray. New York: Routledge, 1995.

———. *In Pursuit of Valis: Selections from the Exegesis.* Edited by Lawrence Sutin. Grass Valley, Calif.: Underwood, 1991.

Dippold, Marc K. "Air Occupation: Asking the Right Questions." *Airpower Journal* 11, 4 (winter 1997): 69–84.

Dudley, Graham. "The Political Ecology of Geographical Information in India." Paper presented at the Initiative 19: GIS and Society Specialist Meeting. Sponsored by the National Science Foundation, South Haven, Minn., March 2–5, 1996.

Duerden, Frank, and Richard Kuhn. "Indigenous Land-Use Information Project." Paper presented at the School of Applied Geography, Ryerson Polytechnic

University, Toronto, 1996, *http://www.ryerson.ca/geog/wwwsag/html/indigen.html,* accessed October 2001.

Duerden, Frank, and Valerie A. Johnson. "GIS and the Visualization of First Nations Land Selections." In *GIS '93 Symposium Proceedings,* 727–730. Vancouver, B.C.: Polaris Conferences, 1993.

Dundas, L. J. L. *India: A Bird's Eye View.* London: Constable, 1924.

Earhart, Amelia. *The Fun of It: Random Records of My Own Flying and of Women in Aviation.* New York: Harcourt Brace, 1932.

Egerton, George. *Keynotes.* London: Elkin Mathews and John Lane, 1893.

Emme, Eugene M. *Aeronautics and Astronautics: An American Chronology of Science and Technology in the Exploration of Space, 1915–1960.* Washington, D.C.: NASA, 1961.

Engel, Marian. *Bear.* Toronto: New Canadian Library, 1991.

Environmental Systems Research Institute, Inc. "Travel and Tourism. Explore the World—ESRI GIS Solutions for the Travel and Tourism Industry." Pamphlet. Redlands, Calif.: Environmental Systems Research Institute, 1998.

Escobar, Arturo. "Imagining a Post-Development Era? Critical Thought, Development, and Social Movements." *Social Text* 30/31 (1992): 20–56.

Fanon, Frantz. *The Wretched of the Earth.* Translated by Constance Farrington. New York: Grove, 1986.

Fawcett, Millicent Garrett. *What I Remember.* Westport, Conn.: Hyperion, 1976.

Fels, John Edward. "Viewshed Simulation and Analysis: An Interactive Approach." *GIS World,* July 1992, 54–59.

Findley, Timothy. *Headhunter.* Toronto: HarperPerennial, 1993.

Foucault, Michel. *The Archaeology of Knowledge and the Discourse of Language.* Translated by A. M. Sheridan Smith. New York: Pantheon, 1972.

———. *Discipline and Punish: The Birth of the Prison.* Translated by Alan Sheridan. New York: Vintage, 1979.

———. "Of Other Spaces." Translated by Jay Miskowiec. *Diacritics* 16 (1986): 22–27.

———. *The Order of Things: An Archaeology of the Human Sciences.* Translated by Alan Sheridan. New York: Vintage, 1973.

Frank, Roman, and David Carruthers. "Networking the Aboriginal Mapping Community in British Columbia." Aboriginal Mapping Network, *http://www.nativemaps.org/,* accessed October 2001.

Freud, Sigmund. "The Uncanny." Translated by Alix Strachey. In *Collected Papers: Papers on Metapsychology: Papers on Applied Psycho-Analysis,* 368–407. Vol. 4. London: Hogarth, 1934.

Freydberg, Elizabeth Hadley. *Bessie Coleman: The Brownskin Lady Bird.* New York: Garland, 1994.

Gabilondo, Joseba. "Postcolonial Cyborgs: Subjectivity in the Age of Cybernetic Reproduction." In *The Cyborg Handbook,* edited by Chris Hables Gray, 423–432. New York: Routledge, 1995.

Gandhi, Leela. *Postcolonial Theory: A Critical Introduction.* New York: Columbia University Press, 1998.

Ghosh, Amitav. *In an Antique Land.* New York: Vintage, 1994.

———. *The Calcutta Chromosome.* New York: Avon, 1995.

Gilroy, Paul. *The Black Atlantic: Modernity and Double Consciousness.* Cambridge: Harvard University Press, 1993.

"*Gis World* Interview—Roger Tomlinson: The Father of GIS." *GIS World,* April 1996, 56–60.

Goddard, George W. *Overview: A Life-Long Adventure in Aerial Photography.* Garden City, N.Y.: Doubleday, 1969.

Gogwilt, Christopher. *The Invention of the West: Joseph Conrad and the Double-Mapping of Europe and Empire.* Stanford: Stanford University Press, 1995.

Gómez, Ricardo, Patrik Hunt, and Emmanuelle Lamoureux. "Telecentros en la Mira: Cómo Pueden contribuir al Desarrollo Social?" [Focus on telecenters: How can they contribute to social development?]. *Chasqui: Latin American Review of Communication,* June 1999, 54–58..

Gore, Al. "The Digital Earth: Understanding Our Planet in the Twenty-First Century." Remarks at the California Science Center, Los Angeles, 1998.

———. "Remarks." Presented at the International Telecommunication Union Plenipotentiary Conference, Minneapolis, October 12, 1998.

Gottmann, Jean. *The Significance of Territory.* Charlottesville: University Press of Virginia, 1973.

"Government Policy on Tibet." Statement from the Foreign and Commonwealth Office, January 1994, *http://www.earthlight.co.nz/users/sonamt/Tibet/TibetFacts11.html,* accessed June 2000.

Gray, Chris Hables. "An Interview with Manfred Clynes." In *The Cyborg Handbook,* edited by Chris Hables Gray, 43–54. New York: Routledge, 1995.

Gray, Chris Hables, *Steven Mentor,* and Heidi J. Figueroa-Sarriera.

"Cyborgology: Constructing the Knowledge of Cybernetic Organisms." In *The Cyborg Handbook,* edited by Chris Hables Gray, 1–14. New York: Routledge, 1995.

Grosz, Elizabeth. *Space, Time, and Perversion.* New York: Routledge, 1995.

Hall, Carol. "Gender and GIS." Paper presented at the Initiative 19: GIS and Society Specialist Meeting. Sponsored by the National Science Foundation, South Haven, Minn., March 2–5, 1996.

Hammond, Rolt. *Air Survey in Economic Development.* New York: American Elsevier, 1967.

Hansen, Desmond. "The Enigma of Lawrence." *Journal of Historical Review* 2, 3: 283–287.

Haraway, Donna. *Simians, Cyborgs, and Women: The Reinvention of Nature.* New York: Routledge, 1991.

Harris, Trevor M., Daniel Weiner, Timothy Warner, and Richard Levin. "Pursuing Social Goals Through Participatory GIS: Redressing South Africa's Historical Political Ecology." In *Ground Truth: The Social Implications of Geographic Information Systems,* edited by John Pickles, 196–222. New York: Guilford Press, 1995.

Harrison, Brian. *Separate Spheres: The Opposition to Women's Suffrage in Britain.* New York: Holmes and Meier, 1978.

Harrowitz, Nancy A. *Antisemitism, Misogyny, and the Logic of Cultural Difference: Cesare Lombroso and Matilde Serao.* Lincoln: University of Nebraska Press, 1994.

Harvey, David. *The Condition of Postmodernity: An Enquiry into the Origins of Cultural Change.* Cambridge: Blackwell, 1990.

Hayler, G. W. "The Airplane and City Planning." *American City* 23, 6 (December 1920): 575–579.

Heiman, Grover. *Aerial Photography: The Story of Aerial Mapping and Reconnaissance.* New York: Macmillan, 1972.

Herodotus, *The Histories.* Translated by A. D. Godley. Loeb Classical Library. 1921.

Holdich, Thomas. "The Use of Practical Geography Illustrated by Recent Frontier Operations." *Geographical Journal* 13, 5 (May 1899): 465–477.

Horkheimer, Max. *The Eclipse of Reason.* New York: Continuum, 1947.

Horkheimer, Max, and Theodor W. Adorno. *Dialectic of Enlightenment.* Translated by John Cumming. New York: Seabury, 1972.

Howse, Derek. *Greenwich Time and the Longitude.* London: Philip Wilson, 1997.

Hunt, Linda. *Secret Agenda: The United States Government, Nazi Scientists, and Project Paperclip, 1945 to 1990.* New York: St. Martin's, 1991.

"The Island of Palmas." Scott, Hague Court Reports 2d 83 (1932) (Perm. Ct. 4rb. 1928), 2 U.N. Rep. Intl. 4rb. Awards 829.

Jacques, T. Carlos. "From Savages and Barbarians to Primitives: Africa, Social Typologies, and History in Eighteenth-Century French Philosophy." *History and Theory* 36, 2 (May 1997): 190–215.

Jameson, Fredric. *Postmodernism: Or, the Cultural Logic of Late Capitalism.* Durham, N.C.: Duke University Press, 1993.

JanMohamed, Abdul. "The Economy of Manichean Allegory: The Function of Racial Difference in Colonialist Literature." *Critical Inquiry* 12, 1 (autumn 1985): 59–87.

Jaros, Dean. *Heroes Without Legacy: American Airwomen, 1912–1944.* Niwot: University Press of Colorado, 1993.

Jennings, R. Y. *The Acquisition of Territory in International Law.* Manchester: Manchester University Press, 1963.

Johnson, B. D. "The Use of Geographic Information Systems by First Nations." School of Community and Regional Planning, University of British Columbia, 1997.

Johnson, Sara E. *To Spread Their Wings: On the Fiftieth Anniversary of the R.C.A.F. (1941–1991).* Spruce Grove, Alberta: Saraband Productions, 1990.

Kamiya, Gary. "An Interview with Michael Ondaatje." *Salon,* November 1996, *http://www.salon.com/nov96/ondaatje961118.html,* accessed January 2000.

Keay, John. *Explorers of the Western Himalayas: 1820–1895.* London: John Murray, 1996.

Kern, Stephen. *The Culture of Time and Space.* Cambridge: Harvard University Press, 1983.

Khadduri, Majid. *Modern Libya: A Study in Political Development.* Baltimore: Johns Hopkins University Press, 1963.

Kipling, Rudyard. *Kim.* New York: Penguin, 1984.

Korman, Sharon. *The Right of Conquest: The Acquisition of Territory by Force in International Law and Practice.* Oxford: Oxford University Press, 1996.

Kuhn, Thomas S. *The Structure of Scientific Revolutions.* Chicago: University of Chicago Press, 1970.

LaDuke, Winona. "We Are Still Here: The Five Hundred Years Celebration." *Sojourners,* October 1991, 12–16.

Lannon, Mary. "Margaret Atwood." *Writers Online Magazine* 3, 1 (fall 1998), *http://www.albany.edu/writers-inst/online.html,* accessed October 2001.

Larbi, Hsen. "The Amazigh World Congress." *Amazigh Voice* 4–5, 31 (December 1995–March 1996), 7–12, 38..

Lawrence, T. E. *Seven Pillars of Wisdom: A Triumph.* Harmondsworth, Eng.: Penguin, 1962.

Lewis, Nelson P. "A New Aid in City Planning." *American City* 26, 3 (September 1922): 253–255.

Lindbergh, Anne Morrow. *Bring Me a Unicorn: Diaries and Letters of Anne Morrow Lindbergh, 1922–1928.* New York: Harcourt Brace Jovanovich, 1971.

———. *Hour of Gold, Hour of Lead: Diaries and Letters of Anne Morrow Lindbergh, 1929–1932.* New York: Harcourt Brace Jovanovich, 1973.

Lively, Penelope. *Moon Tiger.* New York: HarperPerennial, 1987.

Lloyd, Anne, and Liz Newell. "Women and Computers." In *Smothered by Invention: Technology in Women's Lives,* edited by Wendy Faulkner and Erik Arnold. London: Pluto, 1985.

Lloyd, Trevor. *Suffragettes International: The World-Wide Campaign for Women's Rights.* London: American Heritage, 1971.

Locke, John. *The Second Treatise of Government.* Edited by Thomas P. Peardon. New York: Liberal Arts Press, 1952.

Lopez, Barry. *Arctic Dreams: Imagination and Desire in a Northern Landscape.* New York: Scribner's, 1986.

Lyotard, Jean-François. *Just Gaming.* Translated by Wlad Godzich. Minneapolis: University of Minnesota Press, 1985.

———. "Lessons in Paganism." Translated by David Macey. In *The Lyotard Reader,* edited by Andrew Benjamin, 122–154. Oxford: Basil Blackwell, 1989.

———. *Libidinal Economy.* Translated by Iain Hamilton Grant. Bloomington: Indiana University Press, 1993,

———. *The Postmodern Condition: A Report on Knowledge.* Translated by Geoff Bennington and Brian Massumi. Manchester: Manchester University Press, 1984.

McClintock, Anne. *Imperial Leather: Race, Gender, and Sexuality in the Imperial Context.* New York: Routledge, 1995.

McHaffie, Patrick H. "Manufacturing Metaphors: Public Cartography, the Market, and Democracy." In *Ground Truth: The Social Implications of Geographic Information Systems,* edited by John Pickles, 113–129. New York: Guilford, 1995.

Madsen, Wayne "Protecting Indigenous Peoples' Privacy from 'Eyes in the Sky.' " In *Proceedings of the Conference on Law and Information Policy for Spatial Databases.* Edited by H. J. Onsrud. Orono, Me: National Center for Geographic Information and Analysis (NCGIA), 1994.

Markham, Beryl. *West with the Night.* Boston: Houghton Mifflin, 1942.

Markham, Clements R. *A Memoir on the Indian Surveys.* London: Allen, 1878.

Marx, Karl. *Capital.* Vol. 1. Translated by Samuel Moore and Edward Aveling; edited by Frederick Engels. Moscow: Foreign Languages Publishing, 1962.

Mattawa, K. *Isma'ilia Eclipse.* New York: Sheep Meadow, 1997.

Mayall, Kevin, G. Brent Hall, and Thomas Seebohm. "Integrating GIS and CAD to Visualize Landscape Change." *GIS World,* September 1994, 46–49.

Miller, Christopher L. *Blank Darkness: Africanist Discourse in French.* Chicago: University of Chicago Press, 1985.

————. "The Postidentitarian Predicament in the Footnotes of A Thousand Plateaus: Nomadology, Anthropology, and Authority." *diacritics* 23, 3 (1993): 6–33.

Montgomerie, Thomas G. "On the Geographical Position of Yarkund and Other Places in Central Asia." *Proceedings of the Royal Geographical Society* 10 (1866), 162–165.

————. "Report of the Trans-Himalayan Explorations during 1867." *Proceedings of the Royal Geographical Society* 13, 3 (April 12, 1869):

————. "Report on the Trans-Himalayan Explorations, in Connexion with the Great Trigonometrical Survey of India, during 1865–7: Route Survey made by Pundit —, from Nepal to Lhasa, and Thence Through the Upper Valley of the Brahmaputra to Its Source." *Proceedings of the Royal Geographical Society* 12, 3 (July 15, 1868): 148–173.

Moodie, Susanna. *Roughing It in the Bush: Life in the Backwoods.* New York: Lovell, 1884.

Moore, John. "Comin' Home: Defining Anarcho-Primitivism." *Green Anarchist,* summer 1995, 7–8.

Moore-Gilbert, Bart. *Postcolonial Theory: Contexts, Practices, Politics.* New York: Verso, 1997.

Morley, L. W. "Remote Sensing Then and Now." Canada Center for Remote Sensing, *http://www.ccrs.nrcan.gc.ca/ccrs/org/history/historye.html,* accessed November 1999.

Muir, Richard. *History from the Air.* London: Michael Joseph, 1983.

Murray, G. W. "Obituary: Ladislas Almásy." *Geographical Journal* 107, 2 (1951): 253–254.

National Remote Sensing Agency (NRSA). *Guidelines to Use Wasteland Maps.* Hyderabad, India: NRSA, Department of Space, 1991.

Nunes, Mark. "Jean Baudrillard in Cyberspace: Internet, Virtuality, and Post-modernity." *Style* 29, 2 (summer 1995): 314–327.

"Obituary: Lieut.-Colonel Patrick Andrew Clayton." *Geographical Journal* 128, 2 (1962): 254–255.

O'Donnel, J. Hugh. "Global GIS Markets." Paper presented at the GIS/GPS Conference '97, Doha, Quatar, March 2–4, 1997.

Olive, C., and D. Carruthers. "Putting TEK into Action: Mapping the Transition." Paper presented at "Bridging Traditional Ecological Knowledge and Ecosystem Science Conference," Flagstaff, Ariz., August 13–15, 1998.

Ondaatje, Michael. *The English Patient.* New York: Vintage International, 1993.

————. *Running in the Family.* New York: Vintage, 1993.

Paden, W. E. "Theaters of Humility and Suspicion: Desert Saints and New England Puritans." In *Technologies of the Self: A Seminar with Michel Foucault,* edited by L. H. Martin, H. Gutman, and P. H. Hutton, 64–79. London: Tavistock, 1988.

Paffard, Mark. *Kipling's Indian Fiction.* New York: St. Martin's, 1989.

Pankhurst, Christabel. "The Political Outlook." In *Speeches and Trials of the Militant Suffragettes: The Women's Social and Political Union, 1903–1918,* edited by Cheryl R. Jorgensen-Earp. London: Associated University Presses, 1999.

————. *Unshackled: The Story of How We Won the Vote.* London: Hutchinson, 1959.

————. "We Revert to a State of War." In *Speeches and Trials of the Militant Suffragettes: The Women's Social and Political Union, 1903–1918*, edited by Cheryl R. Jorgensen-Earp, London: Associated University Presses, 1999.

Pankhurst, Emmeline. "Address at Hartford." In *Speeches and Trials of the Militant Suffragettes: The Women's Social and Political Union, 1903–1918*, edited by Cheryl R. Jorgensen-Earp. London: Associated University Presses, 1999.

————. "The Argument of the Broken Pane." In *Speeches and Trials of the Militant Suffragettes: The Women's Social and Political Union, 1903–1918*, edited by Cheryl R. Jorgensen-Earp. London: Associated University Presses, 1999.

————. "Great Meeting in the Albert Hall." In *Speeches and Trials of the Militant Suffragettes: The Women's Social and Political Union, 1903–1918*, edited by Cheryl R. Jorgensen-Earp. London: Associated University Presses, 1999.

————. "Kill Me, or Give Me My Freedom!" In *Speeches and Trials of the Militant Suffragettes: The Women's Social and Political Union, 1903–1918*, edited by Cheryl R. Jorgensen-Earp. London: Associated University Presses, 1999.

————. "The Women's Insurrection." In *Speeches and Trials of the Militant Suffragettes: The Women's Social and Political Union, 1903–1918*, edited by Cheryl R. Jorgensen-Earp. London: Associated University Presses, 1999.

Parsons, David W. "British Air Control: A Model for the Application of Air Power in Low-Intensity Conflict?" *Airpower Journal* 8, 2 (summer 1994): 27–39.

Penderel, H.W.G.J. "The Gilf Kebir: A Paper Read with the Following [Richard A. Bermann, "Historic Problems of the Libyan Desert"] at the Evening Meeting of the Society on 8 January 1934." *Geographical Journal* 83, 6 (1934): 450–463.

Perlman, Fredy. *Against History, Against Leviathan*. Detroit: Black and Red, 1983.

Pick, M. "Aerial Photography." *Garden Cities and Town Planning Magazine* 10, 4 (April 1920): 71–86.

Pickles, John. "Representations in an Electronic Age: Geography, GIS, and Democracy." In *Ground Truth: The Social Implications of Geographic Information Systems,* edited by John Pickles, 1–30. New York: Guilford, 1995.

————. "Towards an Economy of Electronic Representation and the Virtual Sign." In *Ground Truth: The Social Implications of Geographic Information Systems,* edited by John Pickles, 223–230. New York: Guilford, 1995.

————, ed. *Ground Truth: The Social Implications of Geographic Information Systems.* New York: Guilford, 1995.

Poole, P. "Indigenous Lands and Power Mapping in the Americas: Merging Technologies." *Native Americas* 15, 4 (winter 1998): 34–43.

Prakrti. "The Breach: Three Sri Lankan-Born Writers at the Crossroads." *Lanka Outlook,* summer 1997, 40–41.

Pratt, Mary Louise. *Imperial Eyes: Travel Writing and Transculturation*. New York: Routledge, 1992.

Pringle, David, and Andy Robertson. "Bayley J. Barrington Interview." *Interzone* 35 (May 1990): 17–21.

Render, Shirley. *No Place for a Lady: The Story of Canadian Women Pilots, 1928–1992*. Winnipeg: Portage and Main, 1992.

"Renew the Earthly Paradise." *Green Anarchist,* autumn 1995, 6.

Rich, Doris L. *Queen Bess: Daredevil Aviator*. Washington, D.C.: Smithsonian Institution Press, 1993.

Roberts, Susan M., and Richard H. Schein. "Earth Shattering: Global Imagery and GIS." In *Ground Truth: The Social Implications of Geographic Information Systems,* edited by John Pickles, 171–195. New York: Guilford, 1995.

Robinson, E. Kay. "Kipling in India." In *Kipling: Interviews and Recollections,* vol. 1, edited by Harold Orel, 67–79. Totowa, N.J.: Barnes and Noble, 1983.

Robinson, James M. *The Nag Hammadi Library in English.* San Francisco: Harper and Row, 1977.

Robinson, S. "Telecenters in Mexico: Learning the Hard Way." Paper presented at "Partnerships and Participation in Telecommunications for Rural Development: Exploring What Works and Why Conference," University of Guelph, Guelph, Ontario, October 26–27, 1998.

Roulet, G. "The New Utopia: Communication Technologies." *Telos* 87 (1992): 39–58.

Rousseau, Jean-Jacques. "The Social Contract." Translated by Gerary Hopkins. In *Social Contract: Essays by Locke, Hume, and Rousseau.* Edited by Ernest Barker. London: Oxford University Press, 1960.

Roy, Arundhati. "The Greater Common Good." In *The Cost of Living.* New York: Modern Library, 1999.

Rubinstein, David. *Before the Suffragettes: Women's Emancipation in the 1890s.* New York: St. Martin's, 1986.

Rushdie, Salman. "At the Auction of the Ruby Slippers." In *East, West.* New York: Vintage International, 1994.

Sack, Robert D. "Territorial Bases of Power." In *Political Studies from Spatial Perspectives,* edited by A. D. Burnett and P. J. Taylor. New York: Wiley, 1981.

Said, Edward. *Culture and Imperialism.* New York: Random House, 1993.

St. Irenaeus of Lyons: Against the Heresies. Translated by Dominic J. Unger. Vol. 1, book 1. New York: Paulist Press, 1992.

Sartre, Jean-Paul. Preface to *The Wretched of the Earth,* by Frantz Fanon. Translated by Constance Farrington. New York: Grove, 1986.

Schrott, Raoul, and Michael Farin. Preface to *Schwimmer in der Wüste,* by Lázló Almásy. Innsbruck: Haymon, 1997.

Seaver, George. *Francis Younghusband: Explorer and Mystic.* London: Murray, 1952.

Segré, Claudio G. *Italo Balbo: A Fascist Life.* Berkeley: University of California Press, 1987.

Shiva, Vandana. *Staying Alive: Women, Ecology, and Development.* London: Zed, 1989.

Short, Nicholas M., Sr. "Remote Sensing and Image Interpretation and Analysis." Pamphlet. Code 935, Goddard Space Flight Center, NASA.

"Six Nations Adapts Traditional Beliefs to New Technology with GIS." *ArcNorth News* 2 (2000):6.

Soja, Edward W. *Postmodern Geographies: The Reassertion of Space in Critical Social Theory.* New York: Verso, 1989.

Spurr, David. *The Rhetoric of Empire: Colonial Discourse in Journalism, Travel Writing, and Imperial Administration.* Durham, N.C.: Duke University Press, 1993.

Standley, Terry. "GIS Implementation in Developing Countries: A United Nations Perspective." Paper presented at the GIS/GPS Conference '97, Doha, Quatar, March 2–4, 1997.

Stokes, I. N. Phelps, and Daniel C. Haskell. *American Historical Prints: Early Views of American Cities, etc. From the Phelps Stokes and Other Collections.* New York: New York Public Library, 1932.

Styles, Showell. *The Forbidden Frontiers: The Survey of India from 1765 to 1949.* London: Hamish Hamilton, 1970.

Swift, Graham. *Out of This World.* London: Picador, 1997.

Taylor, Peter J., and Ronald J. Johnston. "Geographic Information Systems and Geography." In *Ground Truth: The Social Implications of Geographic Information Systems,* edited by John Pickles, 51–67. New York: Guilford, 1995.

Thomas, Lowell. *European Skyways: The Story of a Tour of Europe by Airplane.* Boston: Houghton Mifflin, 1927.

Thrower, Norman J. W. *Maps and Civilization: Cartography in Culture and Society.* Chicago: University of Chicago Press, 1996.

Torgovnick, Marianna. *Gone Primitive: Savage Intellects, Modern Lives.* Chicago: University of Chicago Press, 1990.

Török, Zsolt. "Desert Love: László Almásy, the Real English Patient." *Mercator's World* 2, 5 (September–October 1997): 42–46.

Turner, Frederick Jackson. *The Frontier in American History.* New York: Holt, Rinehart and Winston, 1962.

Vale, V. *Modern Primitives.* San Francisco: Re/Search, 1989.

Varadharajan, Asha. *Exotic Parodies: Subjectivity in Adorno, Said, and Spivak.* Minneapolis: University of Minnesota Press, 1995.

Virilio, Paul. *War and Cinema: The Logistics of Perception.* Translated by Patrick Camiller. New York: Verso, 1989.

Viswanathan, Gauri. "Beyond Orientalism: Syncretism and the Politics of Knowledge." *Stanford Humanities Review* 5, 1 (1995): 19–32.

Ward, David. "Surfacing: Separation, Transition, Incorporation." In *Margaret Atwood: Writing and Subjectivity: New Critical Essays,* edited by Colin Nicholson. New York: St. Martin's, 1994.

Ware, Susan. *Still Missing: Amelia Earhart and the Search for Modern Feminism.* New York: Norton, 1993.

Webster, Donovan. "Journey to the Heart of the Sahara." *National Geographic,* March 1999, 4–33.

White, Hayden. *Tropics of Discourse: Essays in Cultural Criticism.* Baltimore: Johns Hopkins University Press, 1978.

Wilford, John Noble. *The Mapmakers: The Story of Great Pioneers in Cartography from Antiquity to the Space Age.* New York: Vintage, 1982.

Williams, Patrick. "Kim and Orientalism." In *Colonial Discourse and Post-Colonial Theory: A Reader,* edited by Patrick Williams and Laura Chrisman, 480–497. New York: Columbia University Press, 1994.

Winchester, Clarence, and F. L. Wills. *Aerial Photography: A Comprehensive Survey of Its Practice and Development.* London: Chapman and Hall, 1928.

Wingate, O. "In Search of Zerzura." *Geographical Journal* 83, 4 (1934): 281–208.

Woolf, Virginia. "Kew Gardens." In *The Virginia Woolf Reader,* edited by Mitchell A. Leaska. San Diego: Harcourt Brace Jovanovich, 1984.

———. "Mr. Bennett and Mrs. Brown." In *The Captain's Death Bed and Other Essays.* New York: Harcourt Brace, 1956.

————. *Mrs. Dalloway.* New York: Harcourt Brace Jovanovich, 1925.

————. *The Voyage Out.* Edited by Elizabeth Heine. London: Hogarth, 1990.

Yates, Frances A. *The Art of Memory.* Chicago: University of Chicago Press, 1966.

Zepetnek, Steven Totosy de. "Michael Ondaatje's *The English Patient,* 'History,' and the Other." *CLCWeb: Comparative Literature and Culture* 1, 4 (1999), <*http://www.arts.ualberta.ca/clcwebjournal/clcweb99-4/contents99 4.html*>.

Index

Adorno, Theodor, 10–11, 146
aerial photography, 5, 13, 15, 65, 67–69,
    77–83, 86–94, 97, 101–102, 112,
    121–134; aerial cameras, 13, 82, 85,
    88, 94, 101–102; aerial flash photog-
    raphy, 89–91
Aerofilms, 65, 92–93
Afghanistan, 42, 50, 56, 58–60, 169;
    Afridi, 42, 56
Afghan War, 50
Africa, 6–7, 16–17, 33, 77, 84, 96–127,
    170–172, 177
Ahmad, Aijaz, 123, 125
air control strategy, 96–97, 126; in Gulf
    War, 126
Algonquin, 135, 143, 152
Almásy, Lászlo, 96, 100–122, 126
Alnwick, Kenneth J., 97
American Indians, 108, 139–140,
    150–152; Chippewa, 139–140;
    Ojibwe, 108, 140; Suquamish, 151;
    Yakima, 181
anarchism, 29–32, 35–36, 39–40, 177–178;
    Autonomie Club (London), 30–31
Arnold, Henry H., 75
Atwood, Margaret, 132–133, 135–145,
    152–153; *The Journals of Susanna
    Moodie,* 138–139; *Surfacing,* 135–138,
    141–143, 145; *Survival,* 132
Australia, 8, 177; as *Terra nullius*, 8

Bagnold, Ralph, 99–102, 108–110,
    112–113, 117–118
Balbo, Italo, 95–99, 103, 126
Ball, George, 132
Barker, Clive, 155
Barthes, Roland, 52
Baudrillard, Jean, 110, 113, 147, 158,
    178–179

Bayley, Barrington, 155
Beaulieu, Victor-Lévy, 132
Bedouins, 104–105, 108–109, 115,
    119–121, 124
Benjamin, Walter, 86, 91–92
Bermann, Richard, 107–108, 110, 115
Best, Steven, 178
Bhabha, Homi, 24, 59, 113, 124
Blanchard, Madame, 70
Bond, Joyce, 76
Bond, Richard, 118
Booth, William, 33
Borges, Jorge Luis, 6
Borley, S.F.H., 5
Bourdin, Martial, 30–31, 39–40
Bourke-White, Margaret, 77–79
Branch, Melville, 86
Britannia, 26–27
British Empire, 24–25, 27, 38, 40, 43,
    55–56, 96, 108; frontier policy of, 43,
    50, 56–57
Broekhuysen, Jan-Willem, 183
Bromilow, C. O., 120
Brown, Margery, 74

Cachagee, Wayne, 148
Canada, 2, 76, 80–82, 123, 131–153,
    177, 181–183; Canada Land
    Inventory, 134–135, 145–146
cannibalism, 140, 143, 153
capitalism, 133, 176, 178–179
Carpenter, Edmund, 179
Carruthers, David, 183
cartography, 6, 8, 13–16, 43–44, 49, 65, 71,
    79, 91, 131, 145–152, 168, 179–184;
    indigenous and postcolonial, 145–152
Cary, Tina, 183
Chicago World's Fair (1933), 95, 98
China, 42, 48, 56, 60, 175

About the Author

Karen Piper teaches postcolonial studies at the University of Missouri–Columbia. She is the recipient of an NEH award and a Huntington fellowship and has published in journals and books such as *Postcolonial-Literatures: Expanding the Canon, Cultural Critique, American Indian Quarterly*, and *Multi-Ethnic Literature of the U.S.*

310
206
6 + 63